T0220138

EXPLORING MATHEMATICS

WITH INTEGRATED SPREADSHEETS IN TEACHER EDUCATION

EXPLORING MATHEMATICS

WITH INTEGRATED SPREADSHEETS IN TEACHER EDUCATION

Sergei Abramovich

State University of New York at Potsdam, USA

World Scientific

NEW JERSEY · LONDON · SINGAPORE · BEIJING · SHANGHAI · HONG KONG · TAIPEI · CHENNAI · TOKYO

Published by

World Scientific Publishing Co. Pte. Ltd.

5 Toh Tuck Link, Singapore 596224

USA office: 27 Warren Street, Suite 401-402, Hackensack, NJ 07601

UK office: 57 Shelton Street, Covent Garden, London WC2H 9HE

Library of Congress Cataloging-in-Publication Data
Abramovich, Sergei.
 Exploring mathematics with integrated spreadsheets in teacher education / by Sergei Abramovich
(State University of New York at Potsdam, USA).
  pages cm
 Includes bibliographical references and index.
 ISBN 978-9814678223 (hardcover : alk. paper) -- ISBN 978-9814689908 (pbk : alk. paper)
 1. Mathematics--Study and teaching--Data processing. 2. Mathematics teachers--Training of.
3. Electronic spreadsheets in education. I. Title.
 QA20.E43A27 2015
 510.285'554--dc23

                                    2015024760

British Library Cataloguing-in-Publication Data
A catalogue record for this book is available from the British Library.

Printed in Singapore

Preface

The book aims at enhancing mathematical and technological preparation of schoolteachers using an electronic (Microsoft Excel®) spreadsheet integrated with *Maple* [Char *et al.*, 1991] and *Wolfram Alpha* developed by Wolfram Research [www.wolframalpha.com] — digital tools capable of sophisticated symbolic computations that Excel does not provide. The content of the book is a combination of mathematical ideas and concepts associated with pre-college problem-solving curriculum and their extensions into more advanced mathematical topics. The book provides prospective and practicing teachers with a foundation for developing a deep understanding of concepts fundamental to the teaching of school mathematics. It also provides the teachers with a technical expertise in designing spreadsheet-based computational environments. Consistent with the current worldwide guidelines for technology-enhanced teacher preparation, the book emphasizes the unity of context, subject matter, and technology as a method of teaching mathematics.

Throughout the book, a number of mathematics education documents developed worldwide are reviewed as appropriate. More specifically, the book addresses a number of recommendations by the Conference Board of the Mathematical Sciences — an umbrella organization consisting of seventeen professional societies in the United States — for the mathematical preparation of teachers. Among the recommendations published in 2001, one can find the following reference to a spreadsheet: "Computers and an associated explosion in the use of quantitative methods in business and science have dramatically increased the mathematical skills needed in many jobs. Facility at creating spreadsheets is becoming required in many entry level positions for high school graduates" [Conference Board of the Mathematical Sciences, 2001, p. 4]. More recently, the Board reinforced the recommendations about technology use, suggesting, "Teachers should become familiar with various software programs and technology platforms, learning how to use them to analyze data, to reduce computational overhead, to build computational models of mathematical objects, and to perform mathematical experiments" [Conference Board of the Mathematical Sciences, 2012, p. 57]. Similarly, in England, it was argued, "While workplace uses of new technologies (such as structuring real data with spreadsheets and using databases and displays) might be learned when required in a particular context, the use of new technologies to advance mathematical knowledge is not embedded in classroom cultures; yet learners' outside lives and sources of knowledge are significantly influenced by current technology" [Advisory Committee on Mathematics Education, 2011, p. 13]. The book attempts to show how using a spreadsheet integrated with commonly available tools of symbolic computations such as *Maple* and *Wolfram Alpha*, the needed skills can be developed, the learning goals can be achieved, and the classroom cultures updated.

The book emphasizes that whereas mathematical activities that can be motivated by and presented within a spreadsheet may be quite significant, the software allows for hiding some of the complexity and formal structure of mathematics involved. This feature is especially important in the context of preparation of mathematics teachers for it enables their true engagement in a rather advanced content without the need to have a rigorous understanding of the content expected from future professional mathematicians. In other words, using a spreadsheet, one can learn how mathematics can be approached initially through a computational experiment rather than, right away, through a less generally appreciated by the modern students of mathematics formal demonstration. At the same time, in the book, whenever possible, the result of a spreadsheet-based computational experiment is followed by its formal demonstration because the latter, as, for instance, Singaporean Secondary Mathematics Syllabuses put it, "helps students make sense of what they learn in mathematics" [Ministry of Education, Singapore, 2006, p. 4].

It should be noted that proficiency of schoolteachers in the use of a spreadsheet as a tool for conceptual development and educative growth becomes a crucial factor in advancing such use with schoolchildren. According to Australian mathematics educators, the appropriate use of technology "can make previously inaccessible mathematics accessible, and enhance the potential for teachers to make mathematics interesting to more students" [National Curriculum Board, 2008, p. 9]. In Japan, computing technology is considered "not only as a tool for teachers to modify and improve their teaching methods but also as a tool for students to … experience the joy of mathematical activities" [Takahashi, *et al.*,

2006, p. 259]. In the words (quoted in [Abramovich, 2000, p. 44]) of a practicing teacher once enrolled in a course taught by the author: *"What has impressed me most [in the use of spreadsheets] is the visual impact and instant extension of a problem. It really carries a punch! As a teacher I can attest that the "wow" factor is a joy in mathematics classroom. I felt this often as a student from computational environments created on a spreadsheet in our class. I can now use a computer to develop and practice mathematical concepts in my own classroom"*. In support of this quote, the book emphasizes the connection of mathematics for teachers with mathematics for teaching in the digital age.

Finally, the book demonstrates that spreadsheets can be effectively integrated with other technological tools, just as technology, in general, has been integrated with paper and pencil mathematics for the last four decades. The representational efficacy of a spreadsheet as a mathematical modeling tool enables this commonly available computer application to be augmented by user-friendly tools that can support symbolic calculations necessary for the grade appropriate extensions of spreadsheet-based mathematical activities. Learning to design computational environments using diverse capability of generic software tools is consistent with expectations for teachers in Canada who "when using spreadsheets ... could supply students with prepared data sets ... so that students' work with the software would be focused on the mathematics related to the data" [Ontario Ministry of Education, 2005a, p. 15]. By getting ready to meet the expectations, teachers can learn how to use mathematical concepts as tools in computing applications. The combination of context, mathematics, and computing enables them to connect mathematical concepts with corresponding situational referents. This process of teachers' conceptual development in mathematics, for which technology serves as an agency, includes several stages: competence in the use of technology, appreciation of the interplay between experiential and theoretical knowledge, ability to connect seemingly unrelated ideas, gaining experience in technology-motivated conjecturing and follow-up formal demonstration. Helping teachers to proceed through these stages contributes to one of the goals of National Curriculum in England that "aims to ensure that all pupils ... reason mathematically by following a line of enquiry, ... developing an argument, justification or proof using mathematical language" [Department for Education, 2013b, p. 1].

The book consists of ten chapters and Appendix comprised one hundred problems. The first chapter introduces basic spreadsheet functions and one-dimensional modeling techniques. It shows how to generate various numeric sequences using a spreadsheet, model solutions to simple algebraic equations, and utilize integrated spreadsheets as support system for worthwhile algebraic generalizations. The use of descriptive names and circular references in spreadsheet formulas is reviewed as appropriate.

The second chapter focuses on using two-dimensional modeling techniques in the construction of addition and multiplication tables. These basic entities of arithmetic are suggested as rich learning environments spanning across topics and concepts of number theory, algebra, and probability. In particular, the chapter emphasizes the power of technologically supported mathematical induction proof — a classic technique of symbolic computation, which, in the context of integrated spreadsheets, can be outsourced to *Wolfram Alpha* (or *Maple*).

The third chapter introduces a spreadsheet as a modeling tool capable of solving problems described by linear Diophantine equations in three and four unknowns. It is shown how the appropriate use of an integrated spreadsheet can assist the teacher of mathematics in extending traditional boundaries of the available curriculum by posing new grade appropriate problems. In the context of problem posing, the notion of a technology-immune/technology-enabled problem is briefly discussed.

The fourth chapter deals with prime numbers demonstrating how two-dimensional modeling techniques can be used to construct the (spreadsheet-based) sieve of Eratosthenes. In the presence of *Wolfram Alpha*, the suggested uses of a spreadsheet are mostly educative, including its use in the context of prime factorization of integers that can be represented as a sum of two squares in more than one way. Nonetheless, through exploring different prime-producing polynomials (both classic and modern), the chapter shows that a spreadsheet is capable of extending the boundaries of purely educational prime number theory applications. The chapter also includes the use of the *Maple*-based mathematical induction proof in investigating a polynomial producing supposedly the longest string of prime numbers known by the time of writing the book.

The fifth chapter, stemming from the use of a spreadsheet as a manipulative-computational environment for exploring percentage problems at the upper elementary level, presents several concepts of number theory as tools in constructing spreadsheets that enable these kinds of explorations. The major pedagogical idea behind those environments is to support a non-authoritative mathematical discourse that offers more than just right-wrong evaluations. As an unexpected extension of the context of percentage problems, the chapter shows how using a spreadsheet one can generate prime numbers through the Euclidean algorithm and connect the latter to Fibonacci numbers.

The sixth chapter deals with Pythagorean triples and their trigonometric generalization stemming from the Law of Cosines. In particular, it is shown how using a spreadsheet one can be guided by a 'more knowledgeable other' towards the discovery of ancient (third century B.C.) formulas through which whole number sides of right triangles can be calculated. Several classic problems of number theory associated with arithmetical properties of right triangles are explored computationally. Different ways of generating Pythagorean (90°) triples and, more generally, γ – triples are presented.

The seventh chapter deals with problems that traditionally belong to what is commonly referred to as recreational mathematics. The focus of mathematical recreations here is the use of a spreadsheet in exploring properties of integers expressed in terms of their digits. Consequently, specific topics include non-decimal positional systems, the Palindromic Number Conjecture, and the so-called converging differences. The chapter aims at developing skills for computational problem solving of recreational problems included in the Appendix.

The eighth chapter shows how one can use geometric problems that can be traced back to the Babylonian mathematics as background for utilizing algebraic inequalities as means of efficacy of spreadsheet modeling. Here, the process of developing skills in conjecturing, proving, and improving inequalities is strongly application-oriented. Different types of proof are considered and the notion of proof assistant technology in teacher education supported by integrated spreadsheets is discussed.

The ninth chapter deals with modeling real-life (though artificially created) problems through which polygonal numbers and their properties can be introduced and explored. Here, polygonal numbers are used also as tools in generating cyclic sequences mentioned in Chapter 1. Historical remarks about the exploration of polygonal numbers are supported by the use of integrated spreadsheets in searching for triangular squares and, in addition, finding the limiting behavior of their ratios.

The tenth chapter shows some less generally known explorations with Fibonacci numbers leading to the notion of Fibonacci sieve of order *k*. Different generalizations of the classic Golden Ratio are introduced. It is demonstrated how one can use the power of the modern day digital technology for carrying out mathematical proofs requiring extensive symbolic computations perhaps not accessible through the traditional paper-and-pencil algebraic transformations. In this context, it is shown how *Wolfram Alpha* can support the final step of a *Maple*-based mathematical induction proof.

Finally, Appendix includes one hundred problems that can be used for individual or group projects in a technology-enhanced mathematics teacher education course emphasizing the power of integrated spreadsheets as means of exploring various topics in school mathematics and their conceptually grade-appropriate extensions. Many problems are follow-up tasks on the activities introduced throughout the book. Some problems are of recreational nature, yet other (technology-immune/technology-enabled) problems are included to provide challenge to those interested in the use of digital technology for mathematical problem solving.

When preparing this book for publication, the author used Microsoft Excel for Mac 2011 (version 14.4.9), version 18.01 of *Maple*, and the basic version of *Wolfram Alpha*. The last tool is a computational engine with apparently a fluid knowledge base, which can be subject to modification/update at any time. While, in general, all programming-related recommendations of the book should remain accurate as the software updates have been implemented, some small changes in the existing knowledge base of *Wolfram Alpha*, syntax of *Maple*, or features of Excel may take place without any notice. So, like any reference to information available in a non-traditional print is accurate as of the day of its on-line retrieval, everything in this book regarding the features of the three software programs should be error-free as of May 2015.

In conclusion, it is with much admiration that I gratefully acknowledge Prof. Kok Khoo Phua's interest in my proposal for a book submitted to *World Scientific* that, with deeply appreciated support and recommendations of anonymous reviewers, resulted in its publication. Thanks are due to Ms. Rok Ting Tan and Ms. Jacqueline Teo for their editorial guidance during my work on the book. I owe particular gratitude to Prof. Gennady A. Leonov for a helpful advice regarding this project and to Prof. Yuri V. Matiyasevich for his assistance with references to Chapter 4 that I was not aware of. Finally, I am under great obligation to my son Leonard Abramovich, a professional graphic designer, for his insightful design of the cover of the book that illustrates both its historical dimension and intended international flavor.

Sergei Abramovich
Potsdam, NY

Contents

Preface   v

Chapter 1
Basic Spreadsheet Functions and Techniques.....................................   1
   1.1   Introduction...   1
   1.2   Natural Numbers and a Spreadsheet   1
   1.3   Increment as a Variable Entity....................................   5
   1.4   Slider-Controlled Variables in a Spreadsheet
        Environment...   7
   1.5   Generating Arithmetic Sequences of a Variable
        Length   8
   1.6   Spreadsheets and the Agent-Consumer-Amplifier
        Framework   9
        1.6.1 Spreadsheet as an agency for using
              mathematics...   9
        1.6.2 Spreadsheet as a consumer of mathematics.............   13
        1.6.3 Spreadsheet as an amplifier of using
              mathematics...   14
   1.7   Using an Integrated Spreadsheet: An Illustration   16
   1.8   From Conjecturing to Formal Reasoning.........................   21
   1.9   Generating Piece-Wise Monotonous Sequence
        with a Cyclic Behavior..   23
   1.10 An Oscillating Sequence with a Cyclic Behavior..............   26

Chapter 2
Explorations with Two-Dimensional Tables   29
   2.1   Introduction...   29
   2.2   Motivating the Construction of the Addition Table..........   30
        2.2.1 Resolving brainteaser   30
        2.2.2 Proof of formula (2.1).............................   31
   2.3   Motivating the Construction of the Multiplication
        Table   35
        2.3.1 Proof of formula (2.3).............................   38
   2.4   Using a Spreadsheet in the Construction of the
        Addition and Multiplication Tables   40

2.5   Constructing a Single Table of Variable Type
         and Size 42
2.6   Divisibility Property of Numbers in
         the Addition and Multiplication Tables 43
         2.6.1 Counting the multiples of two in
                 the addition and multiplication tables 43
                 2.6.1.1 Counting in the addition tables 43
                 2.6.1.2 Counting in the multiplication tables 44
         2.6.2 Finding the sums of multiples of two in an
                 addition table of even size 45
         2.6.3 Proof of formula (2.5) 46
         2.6.4 Finding the sum of multiples of two in
                 an addition table of odd size 47
         2.6.5 Proof of formula (2.7) 49
2.7   The Sum of the Multiples of Three in
         the $n \times n$ Addition Table 51
         2.7.1 Finding formulas for $\alpha(n, 3)$ 51
         2.7.2 Proving formulas (2.8)-(2.10) 55
                 2.7.2.1 Proof of formula (2.8) 55
                 2.7.2.2 Proof of formula (2.9) 56
                 2.7.2.3 Proof of formula (2.10) 56
2.8   The Sum of the Multiples of Two in
         the $n \times n$ Multiplication Table 57
         2.8.1 The case $n = 2k$ 57
         2.8.2 The case $n = 2k - 1$ 59
         2.8.3 Proof of formula (2.11) 60
         2.8.4 Proof of formula (2.12) 61
2.9   The Sum of the Multiples of Three in
         the $n \times n$ Multiplication Table 62
         2.9.1 The case $n = 3k$ 62
         2.9.2 The case $n = 3k + 1$ 64
         2.9.3 The case $n = 3k - 1$ 65
         2.9.4 Proof of formula (2.13) 66

Chapter 3
Spreadsheet as a 3-D Modeling Tool 69
3.1   Introduction 69

3.2   Spreadsheet Modeling of a Diophantine Equation
       with Three Unknowns 70
3.3   3-D Spreadsheet Modeling Involving Parameters 73
3.4   Using 3-D Spreadsheet Modeling in Problem Posing 77
       3.4.1 A notion of problem coherence 77
       3.4.2 Modeling ambiguous problems using
             a spreadsheet 79
       3.4.3 Using spreadsheets to pose problems 82
       3.4.4 Formulating TITE problems through
             parameterization 83
3.5   Extension into 4-D Modeling 85

Chapter 4
Prime Numbers 87
4.1   Introduction 87
4.2   Generating Prime Numbers Within
       a Spreadsheet 89
4.3   Construction of the Spreadsheet Sieve 91
4.4   Twin Primes 95
4.5   Exploring Gaps Between Consecutive Primes 97
4.6   Testing Prime-Producing Polynomials Using
       a Spreadsheet 98
4.7   Fermat Primes and Euler's Factorization Method 103
       4.7.1 A numeric example 104
       4.7.2 Generalization 104
       4.7.3 Spreadsheet implementation of Euler's
             factorization method 105

Chapter 5
Mathematical Concepts as Modeling Tools 109
5.1   Introduction 109
5.2   Arithmetical Properties of Rectangular Grids 110
5.3   Exploring Syntactic Versatility of a Spreadsheet 112
5.4   From Data-Driven Conjecturing to Proving 115
5.5   Emergence of a New Concept Through
       De-Contextualization of Mediational Means 117

5.5.1 Subgrids of the second order 117
5.5.2 "Real" and "hypothetical" subgrids 118
5.6   The Sum of the Greatest Common Divisors
       as a Problem-Solving Tool 120
5.7   Revisiting Familiar Concepts in New Environments 123
5.7.1 Improving computing complexity 123
5.7.2 Linking the Euclidean algorithm
       to Fibonacci numbers 124
5.7.3 Generating prime numbers through the
       Euclidean algorithm 128

Chapter 6
Pythagorean Triples 133
6.1   Introduction 133
6.2   Towards Finding the General Solution
      to the Pythagorean Equation 135
6.3   Pythagorean Triples Emerge 138
6.4   Computer-Mediated Discourse on
      the Pythagorean Equation 140
6.5   Achieving the Goal and Going Beyond It 141
6.6   Discovering Arithmetical Properties of Pythagorean
      Triangles 144
6.7   Visualizing the Meaning of Fermat's Last Theorem 146
6.8   Pythagorean Triples Meet Euler Phi Function 147
6.9   Generating Pythagorean Triples via the Law of Cosines ... 150
6.10 Technology-Motivated Transition to γ–triples 151

Chapter 7
Recreational Mathematics 157
7.1   Introduction 157
7.2   A Real-Life Context for Recreational Mathematics 158
7.3   Measurement Model for Division as
      a Tool in Computing Applications 160
7.4   A Numeric Example 161
7.5   The Case of an n-Digit Number 163
7.6   Computational Separation of Digits Within
      a Spreadsheet 165

7.7   Illustration I: Resolving the Odometer Challenge 168
7.8   Illustration II: Palindromic Number Conjecture 171
7.9   Converging Differences 172
7.10   Exploring the Growth of the Number of
       Digits in Integer Exponents.. 173

Chapter 8
Historical Connections and Geometric Modeling............................ 177
8.1   Introduction.. 177
8.2   Modeling as Problem Solving in
       a Historical Context 180
8.3   Spreadsheet as an Agent of a Mathematical
       Activity Leading to Inequalities.. 181
8.4   Developing Skills in Conjecturing, Proving,
       and Improving Inequalities 183
8.5   Technology-Motivated Conjecturing................................. 188
8.6   Spreadsheet Modeling vs. Using
       AM-GM Inequality 190
8.7   Making Connections as a Good Teaching Practice........... 196

Chapter 9
Computational Problem Solving in Context 199
9.1   Introduction.. 199
9.2   Establishing Context for Triangular Numbers 201
9.3   Using a Spreadsheet to Investigate Brain Teaser 9.1 202
9.4   Developing Formulas for Triangular Numbers.................. 205
9.5   Visualizing Mathematical Induction Proof
       of Formula (9.4).. 207
9.6   Square Test for Triangular Numbers and
       Its Spreadsheet Implementation 208
9.7   Developing Square Numbers in Context............................ 210
9.8   Investigation of Brain Teaser 9.2 Using
       a Spreadsheet   210
9.9   Developing Formulas for Square Numbers 211
9.10   Using Triangular Numbers as
       Problem-Solving Tools 212
9.11   Two General Formulas for Polygonal Numbers 214

9.12   Square Test for Polygonal Numbers 216
9.13   Triangular-Like Numbers 217
9.14   Inequalities and Search for Triangular-Like
       Numbers Using a Spreadsheet 218
9.15   Square-Like Numbers 221
9.16   Two Types of Ghosts on the Same Floor 222
9.17   Exploring *m*-Gonal-Like Numbers 224
9.18   Two Types of Ghosts Meeting Each Other 226

Chapter 10
Advanced Explorations with Fibonacci Numbers............................. 231
10.1   Introduction.. 231
10.2   Extending Fibonacci Numbers to New Contexts.............. 233
10.3   Fibonacci Sieve of Order *k*.. 236
10.4   From the Golden Ratio to Its Generalization.................... 241
10.5   Generalized Golden Ratios 243
10.6   Computational Experiments with
       Fibonacci-Like Sequences 245

Appendix
One Hundred Problems 251
       A.1   Introduction... 251
       A.2   Problems 252

Bibliography   269

Index   281

Chapter 1

Basic Spreadsheet Functions and Techniques

1.1 Introduction

This chapter introduces basic computational techniques and spreadsheet functions that will be used throughout the book. These techniques include the generation of arithmetic sequences, the use of names as constants and variables in spreadsheet formulas, the use of scroll bars in making a dynamic computational environment depending on parameters controlled by the bars, generating sequences the length of which is controllable through a scroll bar, and the use of circular references in spreadsheet formulas. The functions include logical functions IF, AND, and OR; mathematical functions INT (the greatest integer function) and MOD (remainder function); statistical functions COUNT and COUNTIF; reference functions LOOKUP and INDEX. The chapter shows how a spreadsheet can be used in the context of the agent-consumer-amplifier didactical triad. It provides an illustration of how a spreadsheet can be integrated with other technologies such as *Wolfram Alpha* and *Maple* to allow for the extension of numeric modeling into symbolic computations including the transition from n to $n + 1$ in a mathematical induction proof. The chapter emphasizes the importance of extending conjecturing/generalization motivated by numerical evidence to formal mathematical reasoning. Finally, the chapter shows how one can use a spreadsheet in generating sequences with piece-wise monotonic and cyclic behaviors that are not defined by formula but rather are delineated by a repeated string of numbers forming a pattern.

1.2 Natural Numbers and a Spreadsheet

One of the basic techniques used in spreadsheet modeling is that of generating the natural number sequence 1, 2, 3, 4, ... Throughout the book, these numbers (also called counting numbers) will be used frequently as mediators in developing more complicated numerical sequences or arrays of numbers. Often, a novice user of the software, when faced with the task of generating the first few terms of the sequence of natural numbers, starts typing the numbers one by one into each cell of either column or row of a spreadsheet. Such typing is an automatic process, which is not much different from using paper and pencil. It requires knowing the order of natural numbers, a skill that one develops through pure memorization at the preoperational level of learning arithmetic. Though one acts consciously in typing numbers, his/her attention is directed towards repeating memorized information rather than using a mathematical (recursive) definition of natural numbers. The use of a spreadsheet allows one to become consciously aware of the recursive definition of natural numbers expressed in the form of the difference equation

$$x_{n+1} = x_n + 1, \; n \in N \tag{1.1}$$

subject to the initial condition

$$x_1 = 1. \tag{1.2}$$

In doing so, one learns how a mathematical definition can be embedded into the software to allow for mastering a spreadsheet and using its computational power effectively. Definition (1.1)-(1.2) is a concept that through the use of a spreadsheet brings about conscious awareness of how natural numbers develop. According to Vygotsky [1987], "conscious awareness is an act of consciousness whose object is the activity of consciousness itself" (p. 190); therefore, this act is quite different from automatic repetition of the memorized order of natural numbers. Automatism based on memorization does not create conditions for insight and lacks conscious awareness — an important component for conceptual development. Indeed, "sources of insight can be clogged by automatisms" [Freudenthal, 1983, p. 469]. In a spreadsheet environment, moving away from pure memorization of the order of natural numbers, one can see that each natural number is one greater than the previous number — an operational interpretation of the order of natural numbers. This is how a spreadsheet can be "taught" to remember the order of numbers: the next number is the current number plus one. Whereas human brain does not use addition when naming the number five after the number four, a spreadsheet adds one to four to generate five. Even more interesting, as an alternative to generating five from four a spreadsheet can use knowledge of the (natural) number that immediately precedes four. That is, typing 3 and 4 in two consecutive cells and replicating this pair across a column/row yield the sequence of consecutive natural numbers with the number five being the first term of the sequence.

The recursive development of natural numbers enables one to computerize definition (1.1)-(1.2) by using the so-called ostensive definition. In general, this kind of definition is used to give meaning to an object by *pointing* out instances of this object. Here, the word pointing is the key word for it reduces a possible semantic difficulty to an explicit action. In the context of a spreadsheet, this action includes pointing at the first term, the unity, and defining the second term as the previous term plus one (Fig. 1.1a). In that way, by entering the unity in cell A1 (A1: = 1), defining the formula =A1 + 1 in cell A2 (by pointing/clicking at cell A1), and replicating the formula down the rows, one generates the sequence of consecutive natural numbers (Fig. 1.1b). Alternatively, one can type the numbers 1 and 2 in two consecutive cells and then replicate this pair down the rows. There is an interesting difference in the visually identical outcomes — the spreadsheet hides operation and produces natural numbers one by one as if acting from memorization: there is no formula in a cell that connects two consecutive cells.

	A		A		A		A
1	=1	1	1	1	100	1	100
2	=A1+1	2	2	2	101	2	101
3	=A2+1	3	3	3	102	3	102
4	=A3+1	4	4	4	103	4	103
5	=A4+1	5	5	5	104	5	104
6	=A5+1	6	6	6	105	6	105
7	=A6+1	7	7	7	106	7	106
8	=A7+1	8	8	8	107	8	107
9	=A8+1	9	9	9	108	9	108
10	=A9+1	10	10	10	109	10	109

| (a) | (b) | (c) | (d) |

Fig. 1.1. Formulas (a), numbers (b), new seed value (c), slider-controlled seed value (d).

A spreadsheet enables one not only to generate natural numbers recursively through definition (1.1)-(1.2). Most importantly, it allows one to develop an interactive computational environment that, from the very outset, makes it possible to visualize how various sequences depend on their first term (cell A1, Fig. 1.1b). By changing the entry of cell A1 (Fig. 1.1c) one can see interactively a dynamic change in the sequence. To facilitate the effect of interactivity, cell A1 can be made slider controlled; that is, a slider (called also a scroll bar) can be attached to cell A1 (Fig. 1.1d). Experiencing the development of the sequence of natural numbers as a dynamic process, in which each successive state is the function of the preceding state, creates the image of arithmetic, in particular, and mathematics, in general, as a dynamic subject matter. Formula (1.2) as an element of the formal definition of sequence (1.1) enables one to further the process of developing other number sequences, by altering x_1, within a spreadsheet.

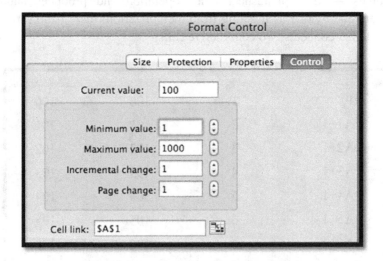

Fig. 1.2. The Format Control dialogue box.

A slider, however, should be specifically created for it is not a part of the standard tool bar. It is located within the tab "Developer" from where it can be dragged to a worksheet and then attached to a cell through the Format Control dialogue box as shown in Fig. 1.2 which relates to cell A1 of the spreadsheet in Fig. 1.1d. Here, both incremental and page changes are set at one. In order to delete a slider or to change data in its dialog box, one has to highlight the slider. To this end, one has to point at the slider and then to press the control button on a (Macintosh) keyboard. This will return the slider to the highlighted state.

1.3 Increment as a Variable Entity

Just as the condition that the first term is equal to one is an element of definition (1.1)-(1.2), there is nothing special about the unity as the increment either. In a spreadsheet environment, an increment, like the first term, may vary. As Fig. 1.3 shows, by entering the numbers 2 and 1 in cells A1 and A2, respectively, the formula

$$=A2+A\$1 \tag{1.3}$$

in cell A3, and replicating this formula down column A, one can interactively generate different number sequences by changing the entries of cells A1 and A2. Note that when one replicates formula (1.3) down column A, the spreadsheet changes a number coordinate of a cell referred to in the first addend and does not change this coordinate in the second addend of formula (1.3). For example, replicating formula (1.3) to cell A10 results in the formula =A9+A\$1. This is due to the \$ sign in formula (1.3) that, in the spreadsheet syntax, designates the coordinate immediately to the right of it to stay the same through the whole process of replication. In other words, formula (1.3), when replicated down column A, assigns no change to cell A1 and it enables change to the first addend beginning from cell A3 down column A. This is reflected in definition (1.1)-(1.2) where x_1 is a constant and x_n changes as n changes. However, x_n does not depend directly on n but rather, it depends on x_{n-1}.

Another way of programming a spreadsheet in order to generate an arithmetic sequence is to use a descriptive name representing the cell within which the increment is defined. This cell can be given the name, say, "inc". To this end, one can highlight cell A1 (Fig. 1.3a), then follow the chain Insert → Name → Define and enter "inc" as a name for the increment into the Define Name dialogue box. Then, the formula =A2 + inc would appear in cell A3, the formula = A3 + inc would appear in cell A4, and so on.

In such a way, the sequences of odd (A1: =2, A2: =1, Fig. 1.3b) and even (A1: =2, A2: =2, Fig. 1.3c) numbers can be generated. (Note that whereas zero may be considered an even number, it does not belong to the set of natural numbers. Therefore, the smallest even number is the number 2). Finally, Fig. 1.3d shows an arithmetic sequence with the first term and difference equal one and three, respectively.

Remark 1.1. The spreadsheet syntax bridges the dynamic perception of the recursive development of odd and even numbers and their description in terms of recursive definition. These descriptions, respectively, are $x_{n+1} = x_n + 2, x_1 = 1, n \in N$ and $x_{n+1} = x_n + 2, x_1 = 2, n \in N$. In fact, these two recursive definitions (alternatively, difference equations) stem from the spreadsheet syntax used in formula (1.3) that introduces a variable (which assumes values of the terms of the sequence) through a relative reference in the first addend and a constant (an increment) through an absolute reference in the second addend. Numerical evidence provided by the spreadsheets pictured in Figs. 1.1b and 1.3b,c suggests that all three sequences (natural, odd, and even numbers) have a constant rate of change; namely, natural numbers have the rate of change equal one, while both odd and even numbers have the rate of change equal two. Because the slope a in the linear function $y = ax + b$ represents its rate of change, closed formulas for the three sequences may be written in the form of linear functions: $x_n = n$ (natural numbers), $x_n = 2n - 1$ (odd numbers), and $x_n = 2n$ (even numbers); $n = 1, 2, 3, \dots$. Both recursive and closed formulas can be implemented within a spreadsheet to generate odd and even numbers. However, unlike recursive formulas, closed formulas, being directly dependent on n, require the use of natural numbers as mediators.

Remark 1.2. A spreadsheet can do more than generating arithmetic sequences (that is, sequences with a constant rate of change). In what follows, arithmetic sequences will be used as mediators in the process of generating numeric sequences with a variable rate of change. That is why, the technique described in this section may be considered as one of the basic techniques used in spreadsheet modeling.

A		A		A		A	
1	=3	1	2	1	2	1	3
2	=1	2	1	2	0	2	1
3	=A2+A$1	3	3	3	2	3	4
4	=A3+A$1	4	5	4	4	4	7
5	=A4+A$1	5	7	5	6	5	10
6	=A5+A$1	6	9	6	8	6	13
7	=A6+A$1	7	11	7	10	7	16
8	=A7+A$1	8	13	8	12	8	19
9	=A8+A$1	9	15	9	14	9	22
10	=A9+A$1	10	17	10	16	10	25

(a)   (b)   (c)   (d)

Fig. 1.3. Spreadsheet as a generator of different arithmetic sequences.

1.4 Slider-Controlled Variables in a Spreadsheet Environment

One of the useful features of a spreadsheet is the possibility of changing the content of a cell automatically by using a slider (alternatively, a scroll bar). Such a slider can be either directly or indirectly linked to a cell the content of which one needs to control, be it a number or a non-numeric text (indirect link). From a technical perspective, the use of a slider makes it possible to change the content of a cell by clicking/dragging rather than typing a new number (or text) in the cell. From a pedagogical perspective, it enables one (especially a younger learner) to interact with a computational environment through a hands-on (informal) approach rather than symbolic (formal) action. In doing so, one grows conceptually for, in the words of historians, *"Like the development of better and stereoscopic vision the growth of manipulative power seems to imply a growth of consciousness. Whenever this happened it was a decisive step toward mastering the world by using it, instead of reacting automatically to it"* [Roberts and Westad, 2013, p. 7]. From a mathematical perspective, a slider adds another conceptual dimension to the environment because a slider-controlled cell can be seen as a variable. For example, one can control the seed value of the natural number sequence (Fig. 1.1c) by attaching a slider to the corresponding cell. In turn, depending on a seed value, a name for this seed value (say, one for the number 1, two for the number 2, and so on) can be generated in another cell through the use of a nested conditional function IF (discussed in detail in the next section).

1.5 Generating Arithmetic Sequences of a Variable Length

A useful technique of spreadsheet programming is the generation of a string of numbers the length of which is a slider-controlled variable. One such example is the generation of different strings of the natural number sequence. To this end, a conditional (logical) function IF can be used. This function has three arguments: a condition to be tested, an action that must be taken if this condition is true, and an action that must be taken if this condition is false. In such a way, if one makes cell A1 a slider-controlled variable, sets the value of this cell at, say, 10, enters number 1 into cell A2, defines the formula =IF(A2<A$1, A2+1, " ") in cell A3, and replicates it down column A, the spreadsheet generates the numbers 1, 2, 3, ..., 10 only; when the value of cell A1 is set at 9, column A displays the numbers 1, 2, 3, ..., 9 only. By "playing" the slider (setting in its Format Control dialog box both the Incremental change and the Page change at 1), one can visualize how the string of natural numbers is expanding or contracting as the number in cell A1 increases or decreases at the click of the slider (Figs. 1.4a, b).

The spreadsheet formula =IF(A2<A$1, A2+1, " ") defined in cell A3 (Fig. 1.4) can be modified to accommodate arithmetic sequences of length n and difference d. To this end, the (algebraic) formula $a_n = a_1 + d(n-1)$ for the n-th term of an arithmetic sequence with the first term a_1 and difference d can be used in spreadsheet programming. Such a spreadsheet is shown in Fig. 1.5 for the (arithmetic) sequence

$$1, 10, 19, 28, ... \tag{1.4}$$

of length ten. In cell A2 the length of the sequence is defined, in cell B2 the difference of the sequence is defined, in cell A3 the first term of the sequence is defined, in cell A4 the formula

=IF(A3<A$3+B$2*(A$2-1), A3+B$2, " ")

is defined and replicated down column A. As a result, after the tenth term, 82, of sequence (1.4) appears in cell A12, all the cells below it stay blank.

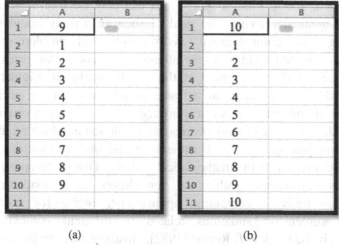

(a)   (b)

Fig. 1.4. A slider-controlled length of a string of natural numbers.

	A	B
1	length	difference
2	10	9
3	1	
4	10	
5	19	
6	28	
7	37	
8	46	
9	55	
10	64	
11	73	
12	82	
13		

Fig. 1.5. Generating sequence (1.4) of length ten.

1.6 Spreadsheets and the Agent-Consumer-Amplifier Framework

1.6.1 *Spreadsheet as an agency for using mathematics*

Many spreadsheet-based mathematical activities can be conceptualized in terms of the agent-consumer-amplifier didactical framework

[Abramovich, 2006]. First, a spreadsheet can play the role of an agent of a mathematical activity. The goal of such an activity could be in developing new computational environments and problem-solving techniques or improving those already available. An important characteristic of this first phase of the activity is that it originates in context, which often serves as a frame of reference for an emerging mathematical concept. A context is understood as a situation originated either inside or outside of mathematics, the resolution of which requires a grade-appropriate application of mathematics. Second, a spreadsheet can become a consumer of the mathematical activity. That is, computational procedures constructed within the spreadsheet using mathematical concepts as tools in computing applications, enable for resolving multiple problematic situations related to different contexts. As mentioned in the Cockcroft Report [1982], "mathematics can be used not only to explain the outcome of an event which has already occurred but also, and perhaps more importantly, to predict the outcome of an event which has yet to take place" (p. 2). Finally, a spreadsheet can be utilized as an amplifier of the mathematical activity in a sense that such utilization makes it possible to solve problems not accessible otherwise; in other words, to amplify the product of the activity.

This didactical triad can be put to work in the context of programming a spreadsheet to enable displaying a string of consecutive natural numbers of the given length. The activity can be illustrated through modeling solutions to a two-variable linear Diophantine equation[a] of the form

$$ax + by = c, \tag{1.5}$$

an indeterminate equation with whole number coefficients a, b, and c that has to be solved in integers. As an example, consider the equation

$$2x + 3y = 20 \tag{1.6}$$

which is a mathematical model of the following problematic situation.

Problem 1.1. *Find all ways to spend $20 at a book sale where books are priced at $2 and $3 only.*

---

[a] The term derives from the name of a third century Greek mathematician Diophantus.

In Eq. (1.6), the variables x and y are the quantities of \$2 and \$3 books, respectively, that can be bought while spending the total of \$20. One way to solve Eq. (1.6) within a spreadsheet is to generate all possible integer values of x followed by the corresponding integer values of y; that is, finding all whole number pairs $(x, \dfrac{20-2x}{3})$. In general, in order to determine the largest possible value of x for which Eq. (1.5) may have a whole number solution, one has to calculate $x_{max} = INT(c/a)$. When $c = 20$ and $a = 2$ we have $x_{max} = 10$. The spreadsheet shown in Fig. 1.6 displays four possible solutions to Eq. (1.6). The following formulas can be used in programming the spreadsheet.

The first formula, =IF(A2<10, A2+1," "), defined in cell A3 and replicated down column A, generates all whole number values of x in the range [0, 10]. It leaves a cell in column A blank when $x > 10$. It is through defining the largest value of x at 10 that the x-range becomes limited to [0, 10]. The second formula,

=IF(A2=" "," ", IF((1/3)*(20-2*A2)=INT((1/3)*(20-2*A2)),(1/3)*
(20-2*A2)," ")),

defined in cell B2 and replicated down column B leaves a cell blank if the corresponding cell in column A is blank (i.e., when $x > 10$); otherwise, the spreadsheet checks whether the corresponding value of y is an integer and if so, this value is displayed, and if not, the cell is left blank. That is, a cell in column B may be blank for two reasons: extraneous range or extraneous solution. This is reflected in the presence of two conditional functions IF in the last formula. In that way, a spreadsheet serves as an agency for a simple mathematical activity including the determination of an exact range within which x varies and finding the corresponding integer values of y within that range. In both cases, the greatest integer (spreadsheet) function INT(x), that returns the largest integer smaller than or equal to x, can be used. Alternatively, one can use an equivalent spreadsheet function FLOOR(x, 1) that rounds x down to the nearest multiple of one, that is, to the nearest integer smaller than or equal to x.

	A	B
1	x	y
2	0	
3	1	6
4	2	
5	3	
6	4	4
7	5	
8	6	
9	7	2
10	8	
11	9	
12	10	0

Fig. 1.6. Four ways of spending $20 for $2 ($x$) and $3 ($y$) books.

One can use a spreadsheet to identify solution pairs of Eq. (1.5) with a specified property. For example, as shown in Fig. 1.7, when $x = y = 4$ such pair is a solution of Eq. (1.6). To this end, one can define the formula =IF(A2=B2, A2, " ") in cell C2 and replicate it down column C. As a result, three fours are displayed back-to-back in row 6. Alternatively, one can use the spreadsheet function LOOKUP which has two parts: the range and the number sought within this range. In that way, as an alternative to the last formula one can define in cell C2 the formula

=IF(A2=" ", " ", IF(B2=LOOKUP(A$2:A$12, A2), A2, " "))

and replicate it down column C. Yet, another spreadsheet function similar to LOOKUP is the function INDEX which has three parts: the range, the horizontal location of the number sought within this range, and its vertical location which is always equal to one in the case when the range is a column. In the case of using the function INDEX, one has to include the sequence of natural numbers as mediators of its second part. Such a spreadsheet is shown in Fig. 1.8 where the natural numbers are included in column A, in columns B and C solutions to Eq. (1.6) are generated, in cell D2 the spreadsheet formula

=IF(B2=" ", " ", IF(C2=INDEX(B$2:B$12, A2, 1), C2," "))

is defined and replicated down column D. Note that applying the function INDEX to an array located within a row rather than a column, the second and the third parts of the function INDEX exchange their places and the natural numbers (mediators) have to be defined in a row rather then in a column. A more sophisticated use of the function INDEX will be demonstrated in Chapter 7.

	A	B	C
1	x	y	
2	0		
3	1	6	
4	2		
5	3		
6	4	4	4
7	5		
8	6		
9	7	2	
10	8		
11	9		
12	10	0	
13			
14			

	A	B	C	D
1		x	y	
2	1	0		
3	2	1	6	
4	3	2		
5	4	3		
6	5	4	4	4
7	6	5		
8	7	6		
9	8	7	2	
10	9	8		
11	10	9		
12	11	10	0	
13				

Fig. 1.7. Using LOOKUP in column C. Fig. 1.8. Using INDEX in column D.

1.6.2 *Spreadsheet as a consumer of mathematics*

Next, the spreadsheet can become a consumer of mathematical activities by considering Eq. (1.5), where a is the price for the object of quantity x, b is the price for the object of quantity y, and c is the total money spent. Such a spreadsheet is shown in Fig. 1.9 in which a, b, and c are slider-controlled variables. The spreadsheet shows that there are three ways to spend \$41 at the \$5 and \$3 book sale.

The following formulas can be used in programming the spreadsheet of Fig. 1.9.

A3: =1; A4: =IF(A3<INT(C\$2/A\$2),A3+1," ")

B3: =IF(A3=" "," ", IF((1/B\$2)*(C\$2-A\$2*A3)=INT((1/B\$2)*
(C\$2-A\$2*A3)), (1/B\$2)*(C\$2-A\$2*A3)," ")).

That is, the inequality A2 < 10 and the expression (1/3)*(20-2*A2) are generalized, respectively, to A3 < INT(C\$2/A\$2) and (1/B\$2)*(C\$2-A\$2*A3).

One can see that a, b and c are slider-controlled variables the change of which enables one to consume mathematical activities that originally were motivated by the use of a spreadsheet. Furthermore, a special technique based on the use of a circular reference — when a spreadsheet formula defined in a cell includes a reference to this cell — enables one to solve the following extension of Problem 1.1.

Problem 1.2. *How many ways can one spend at most $20 buying $3 and $5 books only?*

	A	B	C
1	a	b	c
2	5	3	41
3	1	12	
4	2		
5	3		
6	4	7	
7	5		
8	6		
9	7	2	
10	8		
11			

Fig. 1.9. Three ways of spending $41 on $3 and $5 books.

To answer the last question computationally, the spreadsheet of Fig. 1.9 can be extended to include a two-column table that records the number of possibilities of buying the books when spending at most $20 (Fig. 1.10). The formula

=IF(C$2=1, " ",IF(C$2=D3,COUNT(B$3:B$12), E3))

defined in cell E3 includes reference to cell E3 allowing for the preservation of results in column E (counting the number of possibilities of spending money) obtained at each step of changing the content of the slider-controlled cell C2. In order for a circular reference to be accepted by a spreadsheet formula, one has to open the *Calculation* dialogue box in Excel preferences and check the box *Limit iteration*. One can see (cell E22) that the answer to the question posed in Problem 1.2 is 19.

1.6.3 *Spreadsheet as an amplifier of using mathematics*

At the amplification phase, the spreadsheet of Fig. 1.9 (or Fig. 1.10) can be modified to model solutions to the equation

$$ax^m + by^n = c \qquad (1.7)$$

with whole number coefficients a, b, and c, and the exponents m and n (Fig. 1.11). To this end, the spreadsheet formula defined in cell B3 can be modified to the form

=IF(A3=" ", " ", IF(C$2-A$2*A3^D$2<=0," ",
IF(((1/B$2)*(C$2-A$2*A3^D$2))^(1/E$2)
=INT(((1/B$2)*(C$2-A$2*A3^D$2))^(1/E$2)),
((1/B$2)*(C$2-A$2*A3^D$2))^(1/E$2)," "))).

This formula nests three conditional functions IF. The first condition to be tested is whether the sought value of y belongs to the x-range; the second condition to be tested is whether the expression $c - ax^m$ is non-negative; the third condition is whether the value of y is a whole number. The formula can be simplified (by removing the second condition) if the x-range is refined to take into account the value of the exponent m.

	A	B	C	D	E
1	a	b	c		
2	3	5	20	1	Possibilities
3	0	4		2	0
4	1			3	1
5	2			4	0
6	3			5	1
7	4			6	1
8	5	1		7	0
9	6			8	1
10				9	1
11				10	1
12				11	1
13				12	1
14				13	1
15				14	1
16				15	2
17				16	1
18				17	1
19				18	2
20				19	1
21				20	2
22				Total	19

Fig. 1.10. Computational solution involving circular reference.

Because the above formula uses the range $x^m \in [1, INT(\frac{c}{a})]$, we would then have $x < x^m < INT(\frac{c}{a})$ or $x < \sqrt[m]{INT(\frac{c}{a})}$. That is, cell A3 is entered with zero, cell A4 is entered with the spreadsheet formula =IF(A3<INT(SQRT(C$2/A$2)),A3+1," ") which is replicated down

column A. As a result, the largest value of x generated for the equation $3x^2 + 2y^3 = 66$ (a   special   case   of   Eq.   (1.7)   when $a = 3, b = 2, c = 66, m = 2, n = 3$) is equal to four. This equation has a single whole number solution $x = 2, y = 3$. Note that finding this solution without technology would have required quite an effort on the part of a problem solver. In that way, a spreadsheet can be used to pose and solve non-linear Diophantine equations.

	A	B	C	D	E
1	a	b	c	m	n
2	3	2	66	2	3
3	0				
4	1				
5	2	3			
6	3				
7	4				
8					

Fig. 1.11. Modeling the equation $3x^2 + 2y^3 = 66$.

The following question, however, remains: Is there a context for Eq. (1.7) that can motivate the use of a spreadsheet? To answer this question, note that in general, mathematical models, like Eq. (1.6), originally are motivated by concrete problems, which are to be solved. When tools for solving such models are created (in the digital era these may include computational environments), one often modifies models even before they have any frame of reference. Finding an appropriate real-life situation described by so modified model may come later in the modeling process. In fact, modified models can be used to pose purely mathematical problems solved perhaps by new computational tools.

1.7 Using Integrated Spreadsheets: An Illustration
The agent-consumer-amplifier framework can be applied in the context of integrating a spreadsheet with other software tools to allow for the extension of spreadsheet-based activities in the direction of developing

skills in algebra. Using such powerful tools as *Wolfram Alpha* developed by Wolfram Research and *Maple*, capable of complicated symbolic computations, jointly with a spreadsheet makes it possible for the amplification of numerical modeling activities. As an example, consider sequence (1.4), which is an arithmetic sequence a_n with $a_1 = 1$ and $d = 9$. Therefore, $a_n = a_1 - d(n-1) = 9n - 8$. A task is to find the n-th partial sum of this sequence. One way to find this sum is to use the formula $s_n = \dfrac{a_1 + a_n}{2} n$ to get $s_n = \dfrac{1}{2} n(9n - 7)$. Alternatively, one can enter the first four terms 1, 10, 19, 28, ... into the input box of *Wolfram Alpha* to get $9n - 8$ as the general term of sequence (1.4) and/or enter the command "find the sum $1 + 10 + 19 + 28 + ...$" to get the general expression for its partial sum generated from the sum of the first four terms (Fig. 1.12).

In the context of the so integrated spreadsheet, the environment can further consume this mathematical activity by posing the following (purely algebraic) task decontextualized from its original summation of natural numbers: *Prove that the product $n(9n - 7)$ is divisible by two for any $n \in N$*. Although this task is not difficult to complete by noting that one of the factors in the product is always an even number, using the method of mathematical induction is instructive here for understanding the method which, in the case of cumbersome symbolic computations can be outsourced to technology. The essence of the method was referred to by Pólya [1954] as "the demonstrative phase" (p. 110) or, alternatively, "passing from n to $n + 1$" (p. 111). In demonstrating this transition, one first notes that $n(9n - 7)\big|_{n=1} = 2$, then assumes that the product $n(9n - 7)$ is divisible by two, and finally transforms it as follows $(n+1)(9n+2) = n(9n-7) + 9n + 9n - 7 + 9 = n(9n-7) + 2(9n+1)$. This shows that, starting from $n = 1$, the divisibility by two of the product $n(9n - 7)$ holds true each time n is increased by one, that is, for all $n \in N$.

Input interpretation:

$$1 + 10 + 19 + 28 + \cdots$$

Result:

$$\sum_{n=1}^{\infty}(-8 + 9\,n) \quad \text{(sum does not converge)}$$

Convergence tests:

By the arithmetic series test, the series diverges.

Partial sum formula:

$$\sum_{n=1}^{m}(-8 + 9\,n) = \frac{1}{2}\left(9\,m^2 - 7\,m\right)$$

Fig. 1.12. From a spreadsheet to *Wolfram Alpha*.

$P(n) := n \cdot (9\,n - 7)$

$$n \to n\,(9\,n - 7)$$

$P(n + 1) - P(n)$

$$(n + 1)\,(9\,n + 2) - n\,(9\,n - 7)$$

simplify (%)

$$18\,n + 2$$

Fig. 1.13. *Maple* as an element of an integrated spreadsheet.

In the digital age, the proof can also be carried out in the context of *Maple* as shown in Fig. 1.13 where one defines the function $P(n) = n(9n - 7)$ and then considers the difference $P(n+1) - P(n)$, the simplification of which yields $18n + 2$ — a numerically obvious multiple of two. Therefore, under the assumption that $P(n)$ is divisible by two, the equality $P(n+1) - P(n) = 18n + 2$ implies that $P(n+1)$ is divisible by two as well. One can see how the transition from n to $n + 1$ (or, in the symbolic domain, from $P(n + 1)$ to $P(n)$) can be carried out by computing technology.

One can continue creating partial sums of partial sums in a numeric form within a spreadsheet (Fig. 1.14) where column B includes partial sums $s_n = n(9n-7)/2$ of sequence (1.4) and column C includes partial sums of the (whole number) sequence s_n. Then, entering the command "find the sum $1 + 12 + 42 + 100 + ...$" into the input box of *Wolfram Alpha* yields the n-th partial sum of s_n in the form of the "fraction" $\dfrac{n(n+1)(n+2)(9n-5)}{24}$. This, in turn, leads to a new task:

Prove that the product $n(n+1)(n+2)(9n-5)$ is divisible by 24 for any $n \in N$.

This time, the utilization of the *Maple*-based mathematical induction proof in demonstrating the transition from n to $n + 1$ may be encouraged. As shown in Fig. 1.15, setting $P(n) = n(n+1)(n+2)(9n-5)$ and simplifying the difference $P(n+1) - P(n)$ yields $12(3n+1)(n+2)(n+1)$, an expression explicitly divisible by 12. Yet, the expression's divisibility by 24 is not numerically obvious. To complete the demonstration that $P(n+1) - P(n)$ is divisible by 24, an additional step is needed. One such step could be to show that $(n+1)(n+2)$ is divisible by two. This time, proof can be done without *Maple* by noting that the last expression is the product of two consecutive integers between which one is always an even number. In a similar, but computationally more complex situation, such additional step (or steps) may also be *Maple*-based. For example, in trying to prove with the help of *Maple* that the polynomial $P(n) = n^4 + 6n^3 + 11n^2 + 6n$ is divisible by 24 yields $P(n + 1) - P(n) = 4n^3 + 24n^2 + 44n + 24 = 4(n^3 + 6n^2 + 11n + 6)$, thereby providing a clear demonstration of divisibility of $P(n)$ by four only. An additional step is needed to prove that $n^3 + 6n^2 + 11n + 6$ is divisible by six. This step, unlike the case of proving divisibility by two of the product of two consecutive integers, cannot be articulated in purely qualitative terms without recourse to symbolic computations. One can see that the novel ways of using an integrated spreadsheet for both numeric and symbolic computations are not always straightforward even within apparently identical mathematical contexts. Such an intricacy of human-computer interaction indicates the importance of the notion of pedagogical technological content knowledge [Mishra and Koehler, 2006] as teacher candidates learn mathematical problem solving in the digital era.

	A	B	C	D
1	1	1	1	
2	10	11	12	
3	19	30	42	
4	28	58	100	
5	37	95	195	
6	46	141	336	
7	55	196	532	
8	64	260	792	
9	73	333	1125	
10	82	415	1540	
11	91	506	2046	
12	100	606	2652	

Fig. 1.14. Generating partial sums of partial sums within a spreadsheet.

$P(n) := n \cdot (n+1) \cdot (n+2) \cdot (9n-5)$
$$n \rightarrow n(n+1)(n+2)(9n-5)$$
$P(n+1) - P(n)$
$$(n+1)(n+2)(n+3)(9n+4) - n(n+1)(n+2)(9n-5)$$
$factor(\%)$
$$12(3n+1)(n+2)(n+1)$$

Fig. 1.15. Divisibility by 24 is not numerically obvious.

Remark 1.3. In the absence of numerical evidence, one, lacking the grasp of algebraic symbolism, might have difficulty seeing that the number $n + 2$ immediately follows the number $n + 1$ within a set of natural numbers. For that, one has to see the symbol n as a variable quantity and that adding the number 1 (or 2) to it is a process of increasing this quantity by one (or two), whatever its current value. Here one can observe interplay between thinking about a symbol as a process and seeing it as a concept [Tall *et al.*, 2001].

1.8 From Conjecturing to Formal Reasoning

A useful application of the technique of producing arithmetic sequences of a variable length can be illustrated through a spreadsheet-based solution to the following

Problem 1.3. *Among the terms of arithmetic sequence* (1.4), *how many multiples of two do not exceed a given number?*

In order to solve this problem, one has to generate the strings of sequence (1.4) having a slider-controlled length. (Fig. 1.16 supports all the discussions of this section.) To this end, one has to choose a number, say, 150, enter it into a slider-controlled cell A1, enter 1 into cell A2, define the formula = IF(A2<A$1-9, A2+9, " ") in cell A3, and replicate it down column A. As the spreadsheet pictured in Fig. 16a shows, the largest element of sequence (1.4) that does not exceed 150 is equal to 145. A useful problem-solving practice is to find 145 algebraically after it was found by using a spreadsheet. In doing so, one has to solve the inequality $a_n < 150$ in n where $a_n = 1 + 9(n-1)$ whence $n < 158/9$; the largest value of n satisfying the last inequality is 17; thus $a_{17} = 1 + 9 \cdot 16 = 145$. It is important to distinguish between problems that cannot and can be solved without technology. That is, whereas the above inequality can be solved with the help of, say, *Wolfram Alpha*, such use of an alternative technology tool can only be recommended as a way of verifying the correctness of one's use of algebra.

The next step is to count (electronically) numerical entries in column A that are not greater than 150. To this end, the spreadsheet formula =COUNT(A2:A1000) which counts the number of numerical entries within the range [A2, A1000] can be defined in cell C1 and it returns the number 17. This, however, does not bring an answer to the above question because only even numbers within this string should be counted. How can these numbers be identified by a spreadsheet among the elements of a string?

In order to answer the last question, two spreadsheet functions that will be used throughout the book have to be introduced. These are MOD and COUNTIF — mathematical and statistical functions, respectively. The MOD function has two integer arguments — dividend (D) and divisor (d) — and it returns the remainder after d is divided into

D. The COUNTIF function has two arguments also — range, and an object (either numerical or literal; the latter has to be in quotation marks) to be counted if it appears within that range. In such a way, multiples of two in column A can be identified through the spreadsheet formula =IF(A2=" ", " ", IF(MOD(A2,2)=0, "EVEN", " ")) defined in cell B2 and replicated down column B. As a result, all multiples of two in sequence (1.4) get labeled EVEN. Finally, the spreadsheet formula =COUNTIF(B1:B100, "EVEN") can be defined in cell D1 and it returns the number 8 when cell A1 (the input) contains 150.

By altering the input (the content of cell A1), one can observe change in cells C1 and D1 (Fig. 1.16b). In doing so, one can discover a definitive relationship between numbers displayed in cells C1 and D1; namely, the number of terms in any string of sequence (1.4) is either twice the number of even terms in this string or one more than this number. In other words, $M(l, 2)$ — the number of multiples of two in a string of length l — is either $l/2$ or $(l - 1)/2$. In algebraic form, this observation can be expressed in terms of the function INT defined as the largest integer not exceeding a given number. Therefore, this finding can be presented symbolically as follows: $M(l,2) = INT(l / 2)$. The function INT is a spreadsheet function and one can model the last relationship on a spreadsheet for $l = 1, 2, 3, \ldots$. The following sequence can be generated: 0, 1, 1, 2, 2, 3, 3, 4, 4, ….. . More usages of the function INT will be discussed below.

One can also note that the following subsequence of sequence (1.4) — 10, 28, 46, 64, ... — is an arithmetic sequence with the first term 10 and the difference 18. Thus, one can solve the inequality $10 + 18(n - 1) < 150$ to get $n < 79 / 9$ whence $n = 8$ — the largest whole number solution to the last inequality. Alternatively, knowing that in a numeric sequence of alternating parity there are 17 terms, depending on the first term, there are either 8 or 9 even/odd terms. At the same time, numerical evidence provided by a spreadsheet has great potential to enhance skills in formal mathematical reasoning. Numerical evidence when followed by formal demonstration offers a context that helps one "to understand the difference between ideas that are inductively derived from the results of activities of observation, manipulation and experimentation, and those that are derived deductively" [Takahashi *et al.*, 2006, p. 181].

	A	B	C	D
1	150		17	8
2	1			
3	10	EVEN		
4	19			
5	28	EVEN		
6	37			
7	46	EVEN		
8	55			
9	64	EVEN		
10	73			
11	82	EVEN		
12	91			
13	100	EVEN		
14	109			
15	118	EVEN		
16	127			
17	136	EVEN		
18	145			
19				

	A	B	C	D
1	155		18	9
2	1			
3	10	EVEN		
4	19			
5	28	EVEN		
6	37			
7	46	EVEN		
8	55			
9	64	EVEN		
10	73			
11	82	EVEN		
12	91			
13	100	EVEN		
14	109			
15	118	EVEN		
16	127			
17	136	EVEN		
18	145			
19	154	EVEN		

(a)   (b)

Fig. 1.16(a, b). Identifying and counting specific numbers within a string of numbers.

Remark 1.4. Instead of using the label EVEN, one can instruct the spreadsheet to generate a numerical label, say 1. In this case, one has to replace in the above formulas the part "EVEN" with the number 1. Also, one should keep in mind that not all symbols provided by a spreadsheet could be used as countable labels. The symbol "*" is one such example — it can be used as a marker of a certain property but cannot be counted electronically by using the function COUNT. Likewise, if there is no formula in cell A1, then it is treated as having zero value. Therefore, MOD(A1, 2) = 0 and the formula =IF(A1=" "," ", IF(MOD(A1,2)=0, "EVEN", " ")) acts according to the embedded IF function and labels cell A1 with the word EVEN.

1.9 Generating Piece-Wise Monotonous Sequence with a Cyclic Behavior

Consider the following sequence of numbers

$$1, 2, 3, 4, 1, 2, 3, 4, 1, 2, 3, 4, \ldots \qquad (1.8)$$

in which the string {1, 2, 3, 4} repeats itself over and over. Human ability to generate this and like sequences with a cyclic behavior in a paper-and-pencil environment is automatic and it is based on our understanding that once in the process of counting the largest element of the string (i.e., 4) is reached, we have to start writing the four numbers again, beginning from the smallest number (i.e., 1).

How can sequence (1.8) be generated by a spreadsheet? If one attempts to instruct a spreadsheet to do what a basically educated individual does automatically, one has to describe this process in the formal language of mathematics. What one takes for granted in the automatic process is the skill of counting by one and the conservation of the two endpoints of the string. In the last two sections of this chapter, a spreadsheet will provide a milieu for conceptualizing these essentially automatic skills.

There are two distinct ways of formalizing the process of generating sequence (1.8) and other like sequences with a cyclic behavior. Whereas spreadsheet-based counting by one is made possible by conceptualizing the natural number sequence through recursion, the termination and the resumption of the process of counting can be formally mediated by this sequence. Figure 1.17 shows how consecutive natural numbers defined in column A enumerate sequence (1.8) displayed in column B. By comparing the two columns, one can note that all the numbers (except for the number 4) in column B are just the remainders of the corresponding numbers in column A upon the division by the length of the cycle. In the case of the division by the number 4, this remainder equals zero. In terms of the MOD function, this observation leads to the spreadsheet formula =IF(MOD(A1,4)=0, 4, MOD(A1,4)) defined in cell B1 and replicated down column B. As a result, sequence (1.8) is generated by the spreadsheet (Fig. 1.17).

Sequence (1.8), however, can be generated without using natural numbers as mediators. Indeed, an alternative to mediators is the recursion used for the development of the string {1, 2, 3, 4}. To this end, one has to keep in mind that as soon as the largest value of the string is reached, one has to start counting from the smallest value again, otherwise the process of counting by one continues. In the spreadsheet shown in Fig. 1.17 this idea is expressed by entering the number one as a seed value in cell C1, the spreadsheet formula =IF(C1=4,1,C1+1) in cell C2 and replicating this formula down column C. In addition, both the greatest and the smallest values of the string (i.e., the length of the cycle and its seed value — in Figs. 1.18 and 1.19 cells D1 and E1) as well as the increment of the sequence (in Fig. 1.19 cell F1) could become slider-controlled variables. So, in the spreadsheets of Figs. 1.18 and 1.19 the formula =IF(MOD(A1,D$1)=0, D$1,MOD(A1, D$1)) is defined in cell B1 and replicated down column B yielding a 5-cycle {1, 2, 3, 4, 5} in Fig. 1.18 and a 10-cycle {1, 2, 3, ..., 10} in Fig. 1.19.

	A	B	C
1	1	1	1
2	2	2	2
3	3	3	3
4	4	4	4
5	5	1	1
6	6	2	2
7	7	3	3
8	8	4	4
9	9	1	1
10	10	2	2
11	11	3	3
12	12	4	4
13	13	1	1
14	14	2	2
15	15	3	3
16	16	4	4

Fig. 1.17. Dual modeling of (1.8).

	A	B	C	D	E
1	1	1	7	5	7
2	2	2	8		
3	3	3	9		
4	4	4	10		
5	5	5	11		
6	6	1	7		
7	7	2	8		
8	8	3	9		
9	9	4	10		
10	10	5	11		
11	11	1	7		
12	12	2	8		
13	13	3	9		
14	14	4	10		
15	15	5	11		
16	16	1	7		

Fig. 1.18. Slider-controlled length and seed value.

	A	B	C	D	E	F
1	1	1	3	10	3	5
2	2	2	8			
3	3	3	13			
4	4	4	18			
5	5	5	23			
6	6	6	28			
7	7	7	33			
8	8	8	38			
9	9	9	43			
10	10	10	48			
11	11	1	3			
12	12	2	8			

Fig. 1.19. Generalized cyclic sequences.

Defining (Fig. 1.18) the formulas =E1 in cell C1 and =IF(MOD(A1,D$1)=0, E$1, C1+1) in cell C2 — replicated down column C yields the cycle of length and seed value entered in cells D1 and E1, respectively. In Fig. 1.18 the cycle is {7, 8, 9, 10, 11}. Finally, defining (Fig. 1.19) the formulas =E1 in cell C1 and =IF(MOD(A1,D$1)=0, E$1, C1+F$1) in cell C2 — replicated down column C yields the cycle of length, seed value, and increment entered in cells D1, E1, and F1, respectively. In Fig. 1.19 the cycle is {3, 8, 13, 18, 23, 28}.

1.10 An Oscillating Sequence with a Cyclic Behavior

A more complicated example of a numeric sequence that can be generated within a spreadsheet in two different ways — mediated by the natural number sequence and recursively — is the sequence

$$1, 2, 3, 4, 4, 3, 2, 1, 1, 2, 3, 4, 4, 3, 2, 1, \dots . \qquad (1.9)$$

Sequence (1.9) oscillates with period eight between the largest, 4, and the smallest, 1, values. Whereas counting by one up and down is not difficult by reciting this sequence from memory, a computer has to be instructed in the language of mathematics to do the same electronically. Below, the two formal approaches to such an instruction of a spreadsheet will be described.

To begin the description of the first, recursive approach to generating sequence (1.9) note that in the case of recursion, unlike the case of sequence (1.8), the process of developing sequence (1.9) involves two seed values. That is, at each step, a decision about a number to be generated in a cell of a spreadsheet depends on the content of two cells immediately above it. Furthermore, there are five pivotal points at which a spreadsheet should take action based on a particular behavior of the sequence. This process is described in Fig. 1.20. More specifically, Fig. 1.20(a) shows the situation when an eight-number string {1, 2, 3, 4, 4, 3, 2, 1} is on the rise and its largest value, 4, has been reached. Figure 1.20(b) shows the situation when the string is on the fall and its smallest value, 1, has been reached. Figure 1.20(c) shows the situation when the sequence reaches its smallest value, 1, and stays constant. Figure 1.20(d) shows the situation when the string is on rise and the largest value of the string, 4, has not been reached. Finally, Fig. 1.20(e) shows the situation when a string is on the fall without having reached its smallest value, 1.

The logic presented in the diagram of Fig. 1.20 can be integrated with the spreadsheet syntax as follows. In cells B1 and B2 (Fig. 1.21) the seed values 1 and 2 are defined, respectively; the largest value of the string is defined in cell D1, the nested IF function with function AND imbedded into it

=IF(AND(B2> B1,B2=D$1), B2, IF(AND(B2<B1,B2=1),B2,

IF(AND(B2=B1,B2=1),B2+1,

IF(AND(B2>B1,B2<D$1),B2+1,B2-1))))   (1.10)

is defined in cell B3 and replicated down column B. As a result, column B becomes filled with the elements of sequence (1.9). By making the content of cell D1 a variable; one can generate sequences of the type {1, 2, 3, ..., n, n, n-1, ...2, 1} that oscillate between 1 and n — the smallest and the largest values of the sequence, respectively, — for different values of n.

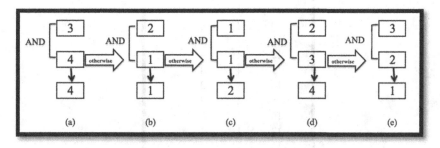

Fig. 1.20. A logic that underlies formula (1.10).

The second approach mediated by the natural number sequence is shown in the spreadsheet of Fig. 1.21. It contains this sequence of numbers in column A and a replica of sequence (1.9) in column C. To clarify the second approach note that each element in the eight-number string {1, 2, 3, 4, 4, 3, 2, 1} can be associated with the remainder from the division of a corresponding natural number by eight (twice the largest number in the string). Indeed, the sequence MOD(n, 8) is a cyclic sequence with period eight; each term of the string (except its last term which always corresponds to a multiple of eight) is either this remainder or one more than the length of the cycle minus the remainder. For example, MOD(7, 8) = 7; then the equality $2 \cdot 4 + 1 - 7 = 2$ relates 7 to 2 in the string with 4 being the largest number. This observation results in the spreadsheet formula

=IF(MOD(A1,2*$D$1)=0,1, IF(MOD(A1, 2*$D$1)

<$D$1+1,MOD(A1,2*$D$1), 2*$D$1+1-MOD(A1,2*$D$1))   (1.11)

defined in cell C1 and replicated down column C. As a result column C becomes filled with the elements of sequence (1.9).

	A	B	C	D
1	1	1	1	4
2	2	2	2	
3	3	3	3	
4	4	4	4	
5	5	4	4	
6	6	3	3	
7	7	2	2	
8	8	1	1	
9	9	1	1	
10	10	2	2	
11	11	3	3	
12	12	4	4	
13	13	4	4	
14	14	3	3	
15	15	2	2	
16	16	1	1	
17	17	1	1	
18	18	2	2	

Fig. 1.21. Generalized spreadsheet for modeling cyclic sequences of type (1.9).

The difference between formula (1.11) and formula (1.10) is that, the former depends on mediators only and, unlike the latter, does not require the use of recursion. Note that formula (1.10) involves 107 characters while formula (1.11) requires 92 characters only. This difference, however, is not significant. Other ways of generating sequence (1.9) within a spreadsheet are possible.

Chapter 2

Explorations with Two-Dimensional Tables

2.1 Introduction

One of the basic uses of a spreadsheet as a two-dimensional modeling tool may deal with the construction of numeric tables. In the context of the elementary school curriculum such tables include addition and multiplication tables. These classic models of arithmetic can be employed for the introduction of many important concepts of school mathematics, such as commutativity, associativity, distributivity, factorization, divisibility, and symmetry. Furthermore, the tables can be used as learning environments for generalization and developing algebraic skills as one is practicing summation techniques by adding either all numbers in the tables or within their special subsets and making the comparison of their similar characteristics. For example, one may compare the sums of all numbers or the sums of the multiples of two in the same size addition and multiplication tables.

As will be demonstrated below, many interesting relationships among whole numbers are implicitly present in the tables and can be revealed through an appropriate pedagogical mediation using a spreadsheet. Included in course work for teacher candidates, these hidden relationships within familiar mathematical structures allow them to "examine mathematics they will teach in depth, from a teacher's perspective" [Conference Board of the Mathematical Sciences, 2012, p. 17] and learn how "when using spreadsheets ... teachers could supply students with prepared data sets ... so that students' work with the software would be focused on the mathematics related to the data" [Ontario Ministry of Education, 2005a, p. 15]. By using addition and multiplication tables, traditionally treated in the schools as mundane learning environments aimed at memorization alone, one can appreciate mathematics as "a subject of enjoyment and excitement, which offers students opportunities for creative work and moments of enlightenment and joy" [Ministry of Education, Singapore, 2006, p. 1]. As a result, one can "become fluent in the fundamentals of mathematics, ... [to allow for] conjecturing relationships and generalisations, and developing an

argument, justification or proof using mathematical language"
[Department for Education, 2013b, p. 1].

2.2 Motivating the Construction of the Addition Table

The construction of the addition table can be motivated by the following

Brain Teaser 2.1: *Within the $n \times n$ checkerboard, find the sum, $\alpha(n)$, of the semi-perimeters of all the types of rectangles.*

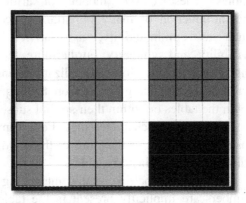

Fig. 2.1. There are nine types of rectangles within a 3×3 checkerboard.

2.2.1 Resolving brainteaser

Let $n = 3$. We have $1 \times 1, 1 \times 2, 1 \times 3, 2 \times 1, 2 \times 2, 2 \times 3, 3 \times 1, 3 \times 2, 3 \times 3$ rectangles (Fig. 2.1) the semi-perimeters of which are equal to, respectively, $2, 3, 4, 3, 4, 5, 4, 5, 6$. The nine integers can be written in the form of a 3×3-matrix $\begin{pmatrix} 2 & 3 & 4 \\ 3 & 4 & 5 \\ 4 & 5 & 6 \end{pmatrix}$ the entries of which are that of the 3×3 addition table shown in Fig. 2.2.

So, by adding all the numbers in the table one gets an answer about the sum of the semi-perimeters. Namely, in the case $n = 3$ (Fig. 2.2) we have

$$\alpha(3) = (2+3+4) + [(2+3+4)+3] + [(2+3+4)+3 \cdot 2]$$
$$= 3 \cdot (2+3+4) + 3 \cdot (1+2) = 36.$$

Likewise, when $n = 4$ (Fig. 2.3) we have

$$\alpha(4) = (2+3+4+5)+[(2+3+4+5)+4]+[(2+3+4+5)+2\cdot4]$$
$$+[(2+3+4+5)+3\cdot4]=4\cdot(2+3+4+5)+4\cdot(1+2+3)=80.$$

	1	2	3
1	2	3	4
2	3	4	5
3	4	5	6

Fig. 2.2. A 3×3 addition table.

	1	2	3	4
1	2	3	4	5
2	3	4	5	6
3	4	5	6	7
4	5	6	7	8

Fig. 2.3. A 4×4 addition table.

Generalizing from the two special cases to the $n\times n$ addition table suggests the following chain of equalities

$$\alpha(n) = (2+3+..+n+1)\cdot n + n\cdot(1+2+3+...+n-1)$$
$$= \frac{n+3}{2}n^2 + n\frac{(n-1)n}{2} = n^2(n+1).$$

That is, the answer to Brain Teaser 2.1 is given by the formula

$$\alpha(n) = n^2(n+1). \tag{2.1}$$

Alternatively, the sum of all numbers in the $n\times n$ addition table is equal to $n^2(n+1)$. That is, within the $n\times n$ checkerboard the sum $\alpha(n)$ of the semi-perimeters of all types of rectangles is equal to $n^2(n+1)$. As mentioned in the Junior High School Teaching Guide for the Japanese Course of Study Mathematics [Takahashi *et al.*, 2006], "students ... can experience the necessity and usefulness [of mathematics] by realizing that there are worlds of quadratic and cubic relationships" (p. 63). By the same token, "validating the conjectures derived from induction deepens our understanding of the content ... [and] is also useful in relating and organizing our knowledge" [*ibid*, p. 59]. This brings about the importance of proof in the teaching of mathematics.

2.2.2 *Proof of formula (2.1)*

Formula (2.1) was conjectured through empirical induction using two special cases. This, however, is not what is sometimes called in mathematics a formal demonstration. Empirical induction is a powerful reasoning tool; yet it has to be complemented by rigorous reasoning afforded by the method of mathematical induction. The full proof by this method consists of two steps: verifying the base clause (typically, the case $n = 1$) and carrying out inductive transfer (i.e., demonstrating that if the statement $P(n)$ is true, the statement $P(n+1)$ is true as well). The use of a spreadsheet made it possible to computationally enhance the second step. Moreover, the conditional formatting feature of the software allows for the inductive transfer to be visually enhanced.

To begin, let $n = 1$. Then $\alpha(1) = 1^2 \cdot (1+1) = 2$ — a true result. Indeed, a 1×1 addition table comprises the number 2 only. Assuming that formula (2.1) holds true for n (in other words, making what can be referred to as inductive assumption), the second step of the demonstrative phase is to show that after replacing n by $n + 1$ it remains true; that is, $\alpha(n+1) = (n+1)^2[(n+1)+1]$. If so, one can pass from $n = 1$ to $n = 2$, then to $n = 3$ and so on, up to infinity, having formula (2.1) satisfied for any value of n. To visually support the second step; in other words, in order to show what augments the table when it acquires new row and new column, one can use conditional formatting. This feature can highlight any augmentation of the table from its current state by a gnomon. As shown in Fig. 2.4, conditional formatting highlights such a gnomon consisting of numbers that belong to $\alpha(10)$ but do not belong to $\alpha(9)$. More specifically,

$$\alpha(10) - \alpha(9) = 2 \cdot (11 + 12 + ... + 20) - 20 = 2 \cdot \frac{11+20}{2} \cdot 10 - 20 = 290.$$

In order to highlight numbers that emerge from the inductive transfer, one has to highlight the range to which Conditional Formatting has to be applied (choose: "Current Selection" in the Manage Rules box, style "Classic", option "Use a formula to determine which cells to format") and then enter the following information into the conditional formatting dialogue box:

Condition 1. Formula is =AND(OR(B$4>n-1, $A5>n-1), COUNT(B$4)>0, COUNT($A5)>0); choose format.

In general, making a transition from n to $n + 1$, as shown in Fig. 2.4 for $n = 9$, we have the following new entries in the $(n+1) \times (n+1)$ addition table:

$$2[(n+2)+(n+3)+...+2(n+1)]-2(n+1)$$

$$=2\frac{(n+2)+2(n+1)}{2}(n+1)-2(n+1)=(n+1)(3n+2).$$

When $n = 9$, we have $(n+1)(3n+2)|_{n=9} = 290$. So, one has to show that

$$\alpha(n+1)=\alpha(n)+(n+1)(3n+2)=(n+1)^2[(n+1)+1].$$

Indeed,

$$n^2(n+1)+(n+1)(3n+2)=(n+1)(n^2+3n+2)$$

$$=(n+1)(n+1)(n+2)=(n+1)^2[(n+1)+1].$$

This completes the proof of formula (2.1).

	A	B	C	D	E	F	G	H	I	J	K	L
1												
2				10								
3												
4			1	2	3	4	5	6	7	8	9	10
5		1	2	3	4	5	6	7	8	9	10	11
6		2	3	4	5	6	7	8	9	10	11	12
7		3	4	5	6	7	8	9	10	11	12	13
8		4	5	6	7	8	9	10	11	12	13	14
9		5	6	7	8	9	10	11	12	13	14	15
10		6	7	8	9	10	11	12	13	14	15	16
11		7	8	9	10	11	12	13	14	15	16	17
12		8	9	10	11	12	13	14	15	16	17	18
13		9	10	11	12	13	14	15	16	17	18	19
14		10	11	12	13	14	15	16	17	18	19	20

Fig. 2.4. Conditional formatting illustrates inductive transfer in proving formula (2.1).

Remark 2.1. Another motivation for the construction of an addition table may also include rolling two dice and computing the sum of the resulting numbers. The 6×6 addition table can explain why having the sum of seven has the highest likelihood, an outcome that second grade students often "are surprised to notice" [Conference Board of the Mathematical Sciences, 2001, p. 92].

Remark 2.2. Common Core State Standards [2010], the modern day educational document in the U.S., set expectations for high school students to "interpret the structure of [algebraic] expressions [and] write expressions in equivalent forms to solve problems" (p. 63). Towards this end, one may note that the product $n^2(n+1)$ can be written as $2n[\dfrac{n(n+1)}{2}]$. Therefore, due to the formula[b] (proved in Sec. 9.5 of Chapter 9)

$$1+2+3+...+n = \frac{n(n+1)}{2}, \qquad (2.2)$$

the sum $\alpha(n)$ can be interpreted as twice the n-multiple of the sum of the first n natural numbers. This connection between the sum of all numbers in the $n \times n$ addition table and the sum of the first n natural numbers can be explained as follows. There are n rows in the table; each row is n greater than the previous row with the first row being n greater than the sum $1+2+3+...+n$. Therefore,

$$\alpha(n) = n(1+2+3+...+n) + n + 2n + 3n + ... + n^2 = 2n(1+2+3+...+n).$$

Remark 2.3. Another way of developing formula (2.1) is to note that the sum of numbers within the gnomon of rank k is equal to $k(3k-1)$ — twice the pentagonal number of that rank, $p_k = \dfrac{k(3k-1)}{2}$ (see Sec. 9.11 of Chapter 9). Indeed, the sum of numbers within the gnomon of rank k is equal to

$$2[(k+1)+(k+2)+...+(k+k)] - 2k = 2 \cdot \frac{3k+1}{2}k - 2k = k(3k-1).$$

Therefore,

$$\alpha(n) = \sum_{k=1}^{n} k(3k-1) = 3\sum_{k=1}^{n} k^2 - \sum_{k=1}^{n} k = 3 \cdot \frac{n(n+1)(2n+1)}{6} - \frac{n(n+1)}{2}$$

$$= n^2(n+1).$$

Here, the formula[c]

---

[b] One of the basic formulas in number theory, by some accounts [Roy, 2011] attributed to Archimedes (the third century B.C, Greece, considered [Rohlin, 2013] among the greatest mathematicians of all times).

[c] According to Kline [1972], "... the sum of the squares of the integers from 1 to 10 was given as though they [Babylonians] have applied the formula $1^2 + 2^2 + ... + n^2 = (1 \cdot 1/3 + n \cdot 2/3)(1+2+...+n)$ " (p. 12).

$$1^2 + 2^2 + ... + n^2 = \frac{n(n+1)(2n+1)}{6}$$

was used.

Remark 2.4. Noting that the $n \times n$ addition table has n^2 entries, formula (2.1) implies that the average entry is equal to $n + 1$. Furthermore, regardless of n, the average entries of the table reside on the bottom left — top right diagonal of the table. In particular, when $n = 10$ the average entry of the corresponding table is $n + 1 = 11$ — this (and only this) number residing in such a diagonal in the spreadsheet of Fig. 2.4.

Remark 2.5. The sum of all numbers in the $n \times n$ addition table can be found by adding numbers residing on the bottom left — top right diagonals equidistant from the top let — bottom right corners of the table. Each such sum is a $2k$ multiple of $n + 1$ — the average entry, where k is the rank of each diagonal in the table, $1 \le k \le n - 1$. For example, in the spreadsheet of Fig. 2.4 the sum $2 + 20$ (adding numbers in cells B5 and K14, respectively) is equal to $22 = 2 \cdot 1 \cdot 11$, thus $k = 1$. Likewise, the sum $(3 + 3) + (19 + 19) = 2 \cdot 2 \cdot 11$, thus $k = 2$. Visualization provided by the spreadsheet-based table of Fig. 2.4 enhances one's thinking about summation strategies in algebraic terms by seeing general in the particular [Mason and Pimm, 1984]. In addition, there are n average entries that the bottom left — top right diagonal include. Therefore,

$$\alpha(n) = \sum_{k=1}^{n-1} 2k(n+1) + n(n+1) = (n+1)[(n-1)n + n] = n^2(n+1).$$

2.3 Motivating the Construction of the Multiplication Table

Brain Teaser 2.2: *Find the number of different rectangles on the $n \times n$ checkerboard.*

As shown in the *Principles and Standards for School Mathematics* [National Council of Teachers of Mathematics, 2000], the total number of rectangles within the $n \times n$ checkerboard can be found by adding up all the numbers in the corresponding $n \times n$ multiplication table. Indeed, when $n = 3$, Fig. 2.1 shows the rectangles

$$1\times1, 1\times2, 1\times3, 2\times1, 2\times2, 2\times3, 3\times1, 3\times2, 3\times3.$$

The number of rectangles of each type on the 3×3 checkerboard is, respectively, 9, 6, 3, 6, 4, 2, 3, 2, and 1. These nine integers can be written in the form of a 3×3-matrix $\begin{pmatrix} 1 & 2 & 3 \\ 2 & 4 & 6 \\ 3 & 6 & 9 \end{pmatrix}$ the entries of which are that of the 3×3 multiplication table shown in Fig. 2.5. This observation leads to the problem of finding the sum of all numbers in the $n\times n$ multiplication table as the total number of rectangles on the 3×3 checkerboard.

	1	2	3
1	1	2	3
2	2	4	6
3	3	6	9

Fig. 2.5. A 3×3 multiplication table.

This sum, $\sigma(n)$, can be found by taking into account that the sum in the *k*-th row of the table is *k* times as big as the sum in the first row. For example (Fig. 2.5), $2+4+6=2\cdot(1+2+3)$ and $3+6+9=3\cdot(1+2+3)$. Therefore, in general,

$$\sigma(n)=(1+2+3+...+n)\sum_{i=1}^{n} i=(1+2+3+...+n)^2=[\frac{n(n+1)}{2}]^2.$$

That is,

$$\sigma(n)=\frac{n^2(n+1)^2}{4}. \qquad (2.3)$$

Recall that

$$1^3+2^3+3^3+...+n^3=\frac{n^2(n+1)^2}{4}.$$

Thus, the sum of all numbers in the $n \times n$ multiplication table equals to the sum $1^3 + 2^3 + 3^3 + \ldots + n^3$ which is the square of the triangular number of rank n. Furthermore, one can see (Fig. 2.5) that $2 + 4 + 2 = 2^3$, $3 + 6 + 9 + 6 + 3 = 3^3$ and (Fig. 2.6) $4 + 8 + 12 + 16 + 12 + 8 + 4 = 4^3$; that is, the sum of numbers within a gnomon is the cube of the gnomon's rank. Alternatively,

$$2^2 + 2^2(2-1) = 2^3, 3^2 + 3^2(3-1) = 3^3, \text{ and } 4^2 + 4^2(4-1) = 4^3.$$

In general, $k^2 + k^2(k-1) = k^3$.

Remark 2.6. One can ask: why do the numbers within a gnomon add up to a cubic number? This question calls for an explanation of the interplay between the procedural perspective on algebraic symbolism and the conceptual understanding of a symbol, k^3, involved. Through such an explanation one can appreciate a link that exists between thinking of symbol as a process and seeing it as a concept [Tall *et al.*, 2001]. To this end, note that the right-bottom element of a k-gnomon, being a symbol representing a square number, can be interpreted as one of the k layers of a $k \times k \times k$-cube which represents a square-based parallelepiped of the unit height. In order to build a cube, one needs to augment this parallelepiped with additional $k - 1$ parallelepipeds of the same size. As was shown above, $k^3 = k^2 \cdot 1 + k^2 \cdot (k-1) = k^2 \cdot 1 + \underbrace{k^2 \cdot 1 + k^2 \cdot 1 + \ldots + k^2 \cdot 1}_{(k-1) \, times}$;

that is, we have a cube consisting of k identical parallelepipeds of the unit height. As Common Core State Standards [2010] put it, "Viewing an expression as the result of operation on simpler expressions can sometimes clarify its underlying structure" (p. 62). In the case of a gnomon, its underlying structure can be interpreted geometrically as building a cube out of parallelepipeds of the unit height. The duality of conceptual perspective on algebraic symbols helps "making visualization a tool for doing and understanding algebra [and] making algebraic manipulation into a tool for geometric understanding, modeling, and proof" [*ibid*, p. 74].

Remark 2.7. Another way of counting the number of rectangles within the $n \times n$ checkerboard is to note that each rectangle has a pair of top-bottom sides (located horizontally) and a pair of left-right sides (located vertically). There are $n + 1$ horizontal lines and $n + 1$ vertical lines; each pair can be chosen in $C_{n+1}^2 = \dfrac{(n+1)!}{2!(n-1)!} = \dfrac{n(n+1)}{2}$ ways. Therefore, there are $\left[\dfrac{n(n+1)}{2}\right]^2$ different rectangles within the $n \times n$ checkerboard. Alternatively, the top line can be put into pairs with other n lines in n ways, the next to the top line can be paired with the remaining $n - 1$ lines in $n - 1$ ways, and so on, until the next to the bottom line is paired with the bottom line in one way. From here, the sum $n + (n-1) + ... + 1$ emerges. Applying formula (2.2) yields the expression $\left[\dfrac{n(n+1)}{2}\right]^2$ as the answer to Brain Teaser 2.2.

2.3.1 *Proof of formula (2.3)*

Formula (2.3) can be proved similarly to formula (2.1) using the method of mathematical induction enhanced visually by a spreadsheet. When $n = 1$ the multiplication table consists of the unity only. By the same token, we have $\sigma(1) = 1$. Just as in the case of the addition table, one should not underestimate the didactical importance of this elementary demonstration for it establishes an important link between the concreteness of the multiplication table and the abstractness of formula (2.3).

Assuming that formula (2.3) holds true for n, one has to show that after replacing n by $n + 1$ it remains true; that is, $\sigma(n+1) = \dfrac{(n+1)^2(n+2)^2}{4}$. To visually support this demonstration; in other words, to show what augments the table when it acquires new row and new column, once again, one can use spreadsheet Conditional Formatting. This feature can highlight any augmentation of the table from its current state by a gnomon. As shown in Fig. 2.6, conditional formatting highlights such a gnomon consisting of numbers that belong to $\sigma(11)$ but do not belong to $\sigma(10)$. More specifically,

$$\sigma(11) - \sigma(10) = 2 \cdot 11 \cdot (1 + 2 + ... + 11) - 11^2.$$

In general, the transition from n to $n + 1$ yields the relationship $\sigma(n+1) = \sigma(n) + 2(n+1)(1 + 2 + ... + (n+1)) - (n+1)^2$. Next, by utilizing formula (2.2) one can show that $\sigma(n+1) = \sigma(n) + (n+1)^3$. Finally, using the inductive assumption about formula (2.3) yields the following chain of equalities

$$\sigma(n+1) = \sigma(n) + (n+1)^3 = \frac{n^2(n+1)^2}{4} + (n+1)^3 = \frac{(n+1)^2(n+2)^2}{4}.$$

This concludes the demonstrative phase of mathematical induction and, thereby, completes the proof of formula (2.3).

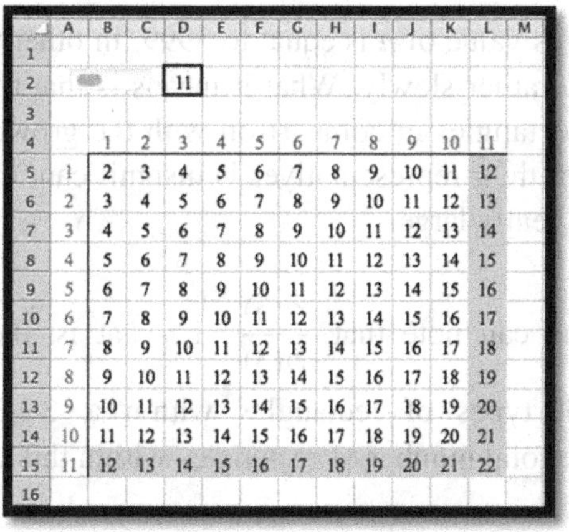

		11									
	1	2	3	4	5	6	7	8	9	10	11
1	2	3	4	5	6	7	8	9	10	11	12
2	3	4	5	6	7	8	9	10	11	12	13
3	4	5	6	7	8	9	10	11	12	13	14
4	5	6	7	8	9	10	11	12	13	14	15
5	6	7	8	9	10	11	12	13	14	15	16
6	7	8	9	10	11	12	13	14	15	16	17
7	8	9	10	11	12	13	14	15	16	17	18
8	9	10	11	12	13	14	15	16	17	18	19
9	10	11	12	13	14	15	16	17	18	19	20
10	11	12	13	14	15	16	17	18	19	20	21
11	12	13	14	15	16	17	18	19	20	21	22

Fig. 2.6. Conditional formatting illustrates inductive transfer in proving formula (2.2).

Remark 2.8. Teacher candidates need experience that later they could impart to their own students in investigating situations involving "quantities that are obtained as a ratio of two quantities of different types" [Takahashi *et al.*, 2004, p. 2004]. The sums $\alpha(n)$ and $\sigma(n)$ are examples of such quantities and, thereby, one can be encouraged to explore the behavior of the ratio $\alpha(n)/\sigma(n)$ using formulas (2.1) and (2.3). To this end, the following relation can be obtained:

$$\frac{\alpha(n)}{\sigma(n)} = \frac{4n^2(n+1)}{n^2(n+1)^2} = \frac{4}{n+1}.$$

Because the fraction $\frac{4}{n+1}$ tends to zero as $n \to \infty$, one can then solve the inequality

$$\frac{4}{n+1} \leq \varepsilon \tag{2.4}$$

as a way of finding the smallest value of n for which the ratio $\dfrac{\alpha(n)}{\sigma(n)}$ becomes smaller than or equal to $\varepsilon > 0$. Solving inequality (2.4) yields $\dfrac{n+1}{4} \geq \dfrac{1}{\varepsilon}$ whence $n \geq \dfrac{4}{\varepsilon} - 1$. Setting $N = \dfrac{4}{\varepsilon} - 1$ one can conclude that for all $n \geq N$ inequality (2.4) holds true. For example, when $\varepsilon = 10^{-2}$, the ratio $\dfrac{\alpha(n)}{\sigma(n)}$ becomes smaller than 10^{-2} only beginning from $n = 399$. When $\varepsilon = 10^{-3}$, this value of n is equal to 3999. In other words, this ratio converges to zero rather slowly. What it means is that the growth of the total number of rectangles in comparison with the growth of the sum of semi-perimeters of their representatives is insignificant unless the size of a checkerboard is really large.

Remark 2.9. One can note that $\dfrac{\alpha(3)}{\sigma(3)} = 1$. That is, the sum of semi-perimeters of all types of rectangles within a 3×3 checkerboard coincides with the total number of rectangles within this checkerboard.

2.4 Using a Spreadsheet in the Construction of the Addition and Multiplication Tables

As shown in the spreadsheet of Fig. 2.7, in the 10×10 addition table column A and row 1 are filled with positive integers 1 through 10. Apparently, entering cell B2 with the spreadsheet formula =B1+A2 results in the number 2. In a similar way, the sums of two integers can be computed in other cells of the table over the range B2:K11.

However, if cell B2 (with the formula =B1+A2) is replicated across all columns (to cell K2), the spreadsheet changes the formula's letter coordinate and preserves its numeric coordinate. By the same token, if cell B2 is replicated down all rows (to cell B11), the spreadsheet changes the formula's numeric coordinate and preserves its letter coordinate. Because the first factor in the formula =B1+A2 relates to row 1 and the second factor relates to column A, one must assign no change to row 1 (relative change to column B), and no change to column A (relative change to row 2). Therefore, if one changes formula in cell B2 to =B$1+$A2 and copies this formula to say, cell E5, the spreadsheet displays number 8 in cell E5 and the formula =E$1+$A5 in the formula bar. (As usual, the $ sign designates the coordinate immediately to the right to stay the same across all columns and down all rows.)

	A	B	C	D	E	F	G	H	I	J	K
1		1	2	3	4	5	6	7	8	9	10
2	1	2	3	4	5	6	7	8	9	10	11
3	2	3	4	5	6	7	8	9	10	11	12
4	3	4	5	6	7	8	9	10	11	12	13
5	4	5	6	7	8	9	10	11	12	13	14
6	5	6	7	8	9	10	11	12	13	14	15
7	6	7	8	9	10	11	12	13	14	15	16
8	7	8	9	10	11	12	13	14	15	16	17
9	8	9	10	11	12	13	14	15	16	17	18
10	9	10	11	12	13	14	15	16	17	18	19
11	10	11	12	13	14	15	16	17	18	19	20

Fig. 2.7. A spreadsheet-based addition table of size ten.

	A	B	C	D	E	F	G	H	I	J	K
1		1	2	3	4	5	6	7	8	9	10
2	1	1	2	3	4	5	6	7	8	9	10
3	2	2	4	6	8	10	12	14	16	18	20
4	3	3	6	9	12	15	18	21	24	27	30
5	4	4	8	12	16	20	24	28	32	36	40
6	5	5	10	15	20	25	30	35	40	45	50
7	6	6	12	18	24	30	36	42	48	54	60
8	7	7	14	21	28	35	42	49	56	63	70
9	8	8	16	24	32	40	48	56	64	72	80
10	9	9	18	27	36	45	54	63	72	81	90
11	10	10	20	30	40	50	60	70	80	90	100

Fig. 2.8. A spreadsheet-based multiplication table of size ten.

Copying cell B2 to cell K11 results in the array of numbers shown in Fig. 2.7. In much the same way, one can construct the 10×10 multiplication table by entering the formula =B$1*$A2 into cell B2 and replicating it to cell K11 (Fig. 2.8). Numbers located on the main (top left — bottom right) diagonal of the multiplication table, as the products of two identical factors, are called square numbers. These numbers will be discussed in Chapter 9.

2.5   Constructing a Single Table of Variable Type and Size

The construction of a table of variable type and size is based on the technique of constructing arithmetic sequences of a variable length (Chapter 1). To this end, the size of a table can be made a slider-controlled variable defined in cell A2 and given the name n (Fig. 2.9). Then, cell B3: = 1; cell A4: = 1; cell C3: =IF(B3<n, 1+B3," ") — replicated across row 3; cell A5: =IF(A4 < n, 1+A4, " ") — replicated down column A; a binary slider with the values 0 and 1 attached to cell O2; cell Q2: =IF(O2=0,"ADD","MULT") — this formula displays the type of the table. Finally, in cell B4 the formula

=IF(OR(B$3=" ",$A4=" ")," ", IF($O$2=1,B$3*$A4,B$3+$A4))

is defined and replicated across rows and down columns of the spreadsheet. This completes the construction of a single table of variable type and size.

	A	B	C	D	E	F	G	H	I	J	K	L	M	N	O	P	Q
1		SIZE													OPERATION		
2	10																ADD
3		1	2	3	4	5	6	7	8	9	10						
4	1	2	3	4	5	6	7	8	9	10	11						
5	2	3	4	5	6	7	8	9	10	11	12						
6	3	4	5	6	7	8	9	10	11	12	13						
7	4	5	6	7	8	9	10	11	12	13	14						
8	5	6	7	8	9	10	11	12	13	14	15						
9	6	7	8	9	10	11	12	13	14	15	16						
10	7	8	9	10	11	12	13	14	15	16	17						
11	8	9	10	11	12	13	14	15	16	17	18						
12	9	10	11	12	13	14	15	16	17	18	19						
13	10	11	12	13	14	15	16	17	18	19	20						
14																	
15																	

Fig. 2.9. A dynamic addition/multiplication table of a variable size.

2.6 Divisibility Properties of Numbers in the Addition and Multiplication Tables

2.6.1 *Counting the multiples of two in the addition and multiplication tables*

How many multiples of two are there in the $n \times n$ addition table? Two cases need to be considered. When n is an even number, such a table is shown in Fig. 2.10 ($n = 10$). When n is an odd number, such a table is shown in Fig. 2.13 ($n = 9$).

How many multiples of two are there in the $n \times n$ multiplication table? Just as in the case of the addition table, two cases need to be considered. When n is an even number, such a table is shown in Fig. 2.21 ($n = 10$). When n is an odd number, such a table is shown in Fig. 2.22 ($n = 9$).

2.6.1.1 *Counting in the addition tables*

In the 10×10 addition table there are 100 entries, half of which are multiples of two (in other words, even numbers). In general, in the $2n \times 2n$ addition table there are $4n^2$ entries, half of which, $2n^2$, are multiples of two. When $n = 5$, we have 50 even sums (Fig. 2.10, cell L2).

In the 9×9 addition table, 32 multiples of two come from the 8×8 table which is augmented by the gnomon of rank nine having nine even entries. In general, in the $(2n+1) \times (2n+1)$ addition table the total number of multiples of two is equal to $2n^2 + (2n+1)$ — half the number of multiples of two in the $2n \times 2n$ table plus the number of multiples of two in the gnomon of rank $2n + 1$. When $n = 4$, we have $32 + 8 + 1 = 41$ even sums (Fig. 2.13, cell L1).

Note that the identity $2n^2 + 2n + 1 = \dfrac{(2n+1)^2 - 1}{2} + 1$ can be interpreted as counting multiples of two in the $(2n+1) \times (2n+1)$ addition table in two ways. The first way is to get the left hand side of the identity as it was mentioned above. The second way is by deleting the sum $4n + 2$ (located in the right-bottom corner) thus making an even number of sums in the table, half of which are multiples of two, and then adding the deleted sum to the total count. In that way, one can learn how to give physical meaning to abstract algebraic identities.

2.6.1.2 *Counting in the multiplication tables*

In the 10×10 multiplication table there are 100 entries three-fourths of which are multiples of two. In general, in the $2n \times 2n$ multiplication table there are $\frac{3}{4} \cdot (2n)^2 = 3n^2$ multiples of two. When $n = 5$, we have 75 even products (Fig. 2.21, cell L1).

In the 9×9 multiplication table there are $3 \cdot 4^2 = 48$ multiples of two that belong to the 8×8 multiplication table plus 8 multiples of two that belong to the gnomon of rank 9. In general, in the $(2n+1) \times (2n+1)$ multiplication table there are $3n^2$ multiples of two that belong to the $2n \times 2n$ table plus $2n$ multiples of two that belong to the gnomon of rank $2n + 1$. In all there are $3n^2 + 2n$ multiples of two in the $(2n+1) \times (2n+1)$ multiplication table. When $n = 4$, we have 56 even products (Fig. 2.22, cell L1).

	A	B	C	D	E	F	G	H	I	J	K	L	M
1		SIZE					DIVISIBLE BY		2			50	
2	10											COUNT	
3		1	2	3	4	5	6	7	8	9	10		
4	1	2		4		6		8		10			
5	2		4		6		8		10		12		
6	3	4		6		8		10		12			
7	4		6		8		10		12		14		
8	5	6		8		10		12		14			
9	6		8		10		12		14		16		
10	7	8		10		12		14		16			
11	8		10		12		14		16		18		
12	9	10		12		14		16		18			
13	10		12		14		16		18		20		
14													
15													
16													
17						550							
18													

Fig. 2.10. Multiples of two in the 10×10 addition table.

2.6.2 Finding the sum of multiples of two in an addition table of even size

One can note that the sums of numbers in each pair of the rows equidistant from the borders of the table are the same, regardless of a pair. More specifically, as shown in Fig. 2.10, the sum of the numbers belonging to the first and the tenth rows can be computed as follows

$$(2 + 4 + 6 + 8 + 10) + (12 + 14 + 16 + 18 + 20)$$

$$= (2 + 20) + (4 + 18) + (6 + 16) + (8 + 14) + (10 + 12)$$

$$= 5 \cdot 22 = 2 \cdot 5 \cdot (10 + 1).$$

Because there are five pairs of rows in the 10×10 addition table (here the number 5 is half the table's size), the sum is equal to $5 \cdot 5 \cdot 22 = 550$ (Fig. 2.10, cell G17).

A similar pattern can be observed in Fig. 2.11, which pictures the 12×12 table. Indeed,

$$(2 + 4 + 6 + 8 + 10 + 12) + (14 + 16 + 18 + 20 + 22 + 24)$$

$$= (2 + 24) + (4 + 22) + (6 + 20) + (8 + 18) + (10 + 16) + (12 + 14)$$

$$= 6 \cdot 26 = 2 \cdot 6 \cdot (12 + 1).$$

Because there are six pairs of rows in the 12×12 addition table (here the number 6 is half the table's size), the sum is equal to $6 \cdot 2 \cdot 6 \cdot 13 = 936$ (Fig. 2.11, cell G17).

One can conjecture from the above two special cases that, in general, the sum of all multiples of two in the $2k \times 2k$ addition table is equal to

$$k \cdot \{(2 + 4 + 6 + ... + 2k) + [(2 + 2k) + (4 + 2k) + (6 + 2k) + ...$$

$$+ (2k + 2k)]\} = k \cdot [(2 + 4k) + (4 + 4k - 2) + (6 + 4k - 4) + ...$$

$$+ (2k + 2k + 2)] = 2 \cdot k^2 \cdot (2k + 1).$$

Setting the sum of all even numbers in the $2k \times 2k$ addition table $\alpha(2k, 2)$, we have the formula

$$\alpha(2k, 2) = 2k^2(2k + 1). \tag{2.5}$$

In particular, when $k = 5$ (10×10 table) and $k = 6$ (12×12 table) we have, respectively, $2k^2(2k + 1)\big|_{k=5} = 2 \cdot 25 \cdot 11 = 550$ and $2k^2(2k + 1)\big|_{k=6} = 2 \cdot 36 \cdot 13 = 936$ — the results confirmed computationally in Figs. 2.10 and 2.11.

	A	B	C	D	E	F	G	H	I	J	K	L	M
1		SIZE				DIVISIBLE BY		2				72	
2	12										COUNT		
3		1	2	3	4	5	6	7	8	9	10	11	12
4	1	2		4		6		8		10		12	
5	2		4		6		8		10		12		14
6	3	4		6		8		10		12		14	
7	4		6		8		10		12		14		16
8	5	6		8		10		12		14		16	
9	6		8		10		12		14		16		18
10	7	8		10		12		14		16		18	
11	8		10		12		14		16		18		20
12	9	10		12		14		16		18		20	
13	10		12		14		16		18		20		22
14	11	12		14		16		18		20		22	
15	12		14		16		18		20		22		24
16													
17						936							
18													

Fig. 2.11. New even entries in the addition table as *n* changes from 10 to 12.

2.6.3 *Proof of formula (2.5)*

Formula (2.5) can be proved by the method of mathematical induction. When $k = 1$ we have $\alpha(2, 2) = 6$. This result is confirmed by adding the numbers 2 and 4, these being the only entries of the addition table of size two. In order to carry out the transition from k to $k + 1$, one can observe the augmentation of the table shown in Fig. 2.11. In the range L4:M15 there are six pairs of numbers with the sum 36. Taking into account that same sum appears in the range B14:M15 and that the two numbers that belong to the overlap of the above two ranges have not to be counted twice, we have $2 \cdot 6 \cdot 36 - 46 = 386$, the result confirmed computationally by comparing cell G17 in Fig. 2.11 to cell G17 in Fig. 2.10. Indeed,

$$\alpha(12, 2) - \alpha(10, 2) = 936 - 550 = 386.$$

Therefore, assuming that formula (2.5) is true for k, one has to show that it remains true when k is replaced by $k + 1$. Because the augmentation of the sum in the transition from k to $k + 1$ can be expressed as

$$2(k+1)[2(k+1) + 2(k+1) + 2k + 2] - (8k+6) = 12(k+1)^2 - (8k+6),$$

one has to prove the following identity

$$2k^2(2k+1) + 12(k+1)^2 - (8k+6) = 2(k+1)^2(2k+3).$$

As shown in Fig. 2.12, *Wolfram Alpha* confirms the identity because the difference between its left and right hand sides is equal to zero.

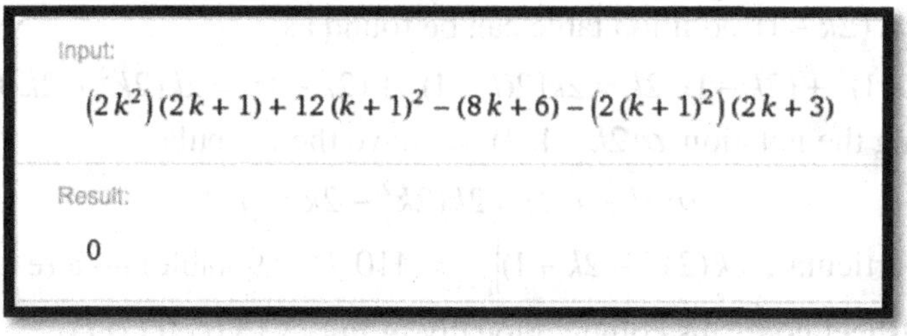

Input:

$$(2 k^2)(2 k + 1) + 12 (k + 1)^2 - (8 k + 6) - (2 (k + 1)^2)(2 k + 3)$$

Result:

0

Fig. 2.12. Using *Wolfram Alpha* in verifying inductive transfer.

Remark 2.10. Using *Wolfram Alpha* in proving formula (2.5) constituted the critical step in supporting symbolic computations in the context of inductive transfer. Such technological support of teacher candidates' doing mathematics provides an examples of their engagement "in the use of a variety of technological tools, including those designed for mathematics ... even if these tools are not the same ones they will eventually use with students" [Conference Board of the Mathematical Sciences, 2012, p. 50].

2.6.4 *Finding the sum of multiples of two in an addition table of odd size*

When *n* is an odd number (e.g., in Fig. 2.13 (cell A2) $n = 9$), the sum of the multiples of two in the $n \times n$ addition table can be found as follows. Note that this time not all rows have the same number of entries. However, if one considers the so-called shallow diagonals, it can be observed that the sums of numbers in the shallow diagonals, which are equidistant from the table's top-left and bottom-right, represent a constant value connected to the size of the table. More specifically, as shown in Fig. 2.13 (cell G17), the sum of all the multiples of two can be found as follows:

$$1 \cdot (2 + 18) + 3 \cdot (4 + 16) + 5 \cdot (6 + 14) + 7 \cdot (8 + 12) + 9 \cdot 10$$

$$= 20 \cdot (1 + 3 + 5 + 7) + 9 \cdot 10 = 20 \cdot 4^2 + 9 \cdot 10 = 410.$$

Note that the equality $1 + 3 + 5 + 7 = 4^2$ used above is a special case of the formula

$$1+3+5+...+2n-1=n^2, \tag{2.6}$$

stating that the sum of the first n odd numbers is equal to n^2.

In general, using formula (2.6), the sum of all multiples of two in the $(2k-1)\times(2k-1)$ addition table can be found as

$$4k(k-1)^2+(2k-1)\cdot 2k=2k[2(k-1)^2+(2k-1)]=2k(2k^2-2k+1).$$

Using the notation $\alpha(2k-1,2)$ we have the formula

$$\alpha(2k-1,2)=2k(2k^2-2k+1). \tag{2.7}$$

In particular, $2k(2k^2-2k+1)\big|_{k=5}=410$ (9×9-table) — a result that a spreadsheet confirms computationally in Fig. 2.13 (cell G17).

	A	B	C	D	E	F	G	H	I	J	K	L	M
1		SIZE				DIVISIBLE BY			2			41	
2	9											COUNT	
3		1	2	3	4	5	6	7	8	9			
4	1	2		4		6		8		10			
5	2		4		6		8		10				
6	3	4		6		8		10		12			
7	4		6		8		10		12				
8	5	6		8		10		12		14			
9	6		8		10		12		14				
10	7	8		10		12		14		16			
11	8		10		12		14		16				
12	9	10		12		14		16		18			
13													
14													
15													
16													
17						410							
18													

Fig. 2.13. Multiples of two in the 9×9 addition table.

Remark 2.11. Some mathematics education standards [e.g., Ministry of Education, Singapore, 2006] suggest that learning opportunities for students "to seek alternative ways of solving the same problem ... develop the metacognitive awareness ... and the ability to control one's thinking process" (p. 5). With this in mind, one can be encouraged to use a different summation strategy in finding the sum of even numbers in an odd size addition table. For example, as shown in Fig. 2.13, adding numbers along the pairs of the diagonals parallel to the main (top left — bottom right) diagonal and equidistant from the bottom-left and top-right corners) yields

$$(20 + 60 + 100 + 140) + 90 = 20 \cdot (1 + 3 + 5 + 7) + 9 \cdot 10$$

$$= 4 \cdot 5 \cdot 4^2 + (2 \cdot 5 - 1) \cdot 2 \cdot 5.$$

Generalizing from the last relation yields formula (2.7). Note that whereas the selection of summation strategy is due to mathematical reasoning skills, numeric and symbolic computations can be outsourced to software tools.

	A	B	C	D	E	F	G	H	I	J	K	L	M
1		SIZE				DIVISIBLE BY		2				61	
2	11										COUNT		
3		1	2	3	4	5	6	7	8	9	10	11	
4	1	2		4		6		8		10		12	
5	2		4		6		8		10		12		
6	3	4		6		8		10		12		14	
7	4		6		8		10		12		14		
8	5	6		8		10		12		14		16	
9	6		8		10		12		14		16		
10	7	8		10		12		14		16		18	
11	8		10		12		14		16		18		
12	9	10		12		14		16		18		20	
13	10		12		14		16		18		20		
14	11	12		14		16		18		20		22	
15													
16													
17							732						
18													

Fig. 2.14. Transition from $n = 9$ to $n = 11$ in the addition table.

2.6.5 Proof of formula (2.7)

Formula (2.7) can be proved by the method of mathematical induction. When $k = 1$ we have $\alpha(1, 2) = 2$. This result is confirmed by Fig. 2.13, as the number 2 is the only entry of the addition table of size one. To demonstrate the transition from k to $k + 1$, one can observe the augmentation shown in Fig. 2.14. In the range K4:L14 there are five pairs of numbers with the sum 32. Taking into account that same sum appears in the range B13:K14 and that one number that belong to the overlap of the above two ranges has not to be counted twice and, in addition, another number two more than the twice counted belongs to the augmentation, we have $2 \cdot 5 \cdot 32 + 2 = 322$, the result confirmed computationally by comparing cell G17 in Fig. 2.14 to cell G17 in Fig. 2.13. Indeed,

$$\alpha(2 \cdot 5 + 1, 2) - \alpha(2 \cdot 5, 2) = 732 - 410 = 322.$$

Therefore, assuming that formula (2.7) is true for k, one has to show that it remains true when k is replaced by $k + 1$. Because the augmentation of the sum in the transition from k to $k + 1$ can be expressed as

$$2[k(2k + 2 + 4k)] + 2 = 12k^2 + 4k + 2,$$

one has to prove the following identity

$$2k(2k^2 - 2k + 1) + 12k^2 + 4k + 2 = 2(k + 1)^2[2(k + 1)^2 - 2(k + 1) + 1].$$

As shown in Fig. 2.15, the identity has been confirmed by *Wolfram Alpha* which yields zero when calculating the difference between its left and right hand sides.

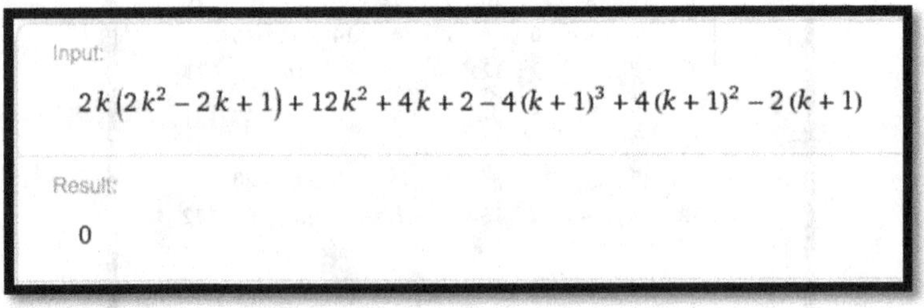

Input:

$$2k(2k^2 - 2k + 1) + 12k^2 + 4k + 2 - 4(k + 1)^3 + 4(k + 1)^2 - 2(k + 1)$$

Result:

0

Fig. 2.15. Confirming inductive transfer by *Wolfram Alpha*.

Remark 2.12. The joint use of a spreadsheet with other software tools that support associated operations like symbolic computations not available in Excel (called an integrated spreadsheet) was demonstrated in Sec. 1.7 of Chapter 1. As shown by Abramovich *et al.* [2014], using integrated spreadsheets that include not-computational tools like the images of modern technologies familiar to young students such as the Nintendo DS, the PlayStation Portable, and the iPhone, made it possible to enhance mathematical activities by the students. Likewise, as symbolic computation activities arise in the context of a spreadsheet, they may be outsourced to other tools capable of supporting those activities. As recommended in the Junior High School Teaching Guide for Japanese Course of Study Mathematics [Takahashi *et al.*, 2006], whereas students should be able to do simple algebraic transformations, they should "avoid using unnecessary complex algebraic expressions whose transformations will not serve any purpose" (p. 171). At the same time, when the purpose of complex transformations is proof, in the digital era, students should learn how to use technology in support of this mathematically significant purpose. Indeed, "ICT tools empower students to work on problems which would otherwise require more advanced mathematics" [Ministry of Education, Singapore, 2012, p. 31].

2.7 The Sum of the Multiples of Three in the $n \times n$ Addition Table

Let $\alpha(n, 3)$ denote the sum of all multiples of 3 in the $n \times n$ addition table. Similarly to the case of finding $\alpha(n, 2)$ when even and odd values of n had to be considered separately, three cases need to be considered now: $n = 3k - 1, n = 3k, n = 3k + 1$.

2.7.1 *Finding formulas for* $\alpha(n, 3)$

Let $n = 3k - 1 = \{2, 5, 8, 11, ...\}$. The case $k = 3$ (i.e., $n = 8$) is shown in Fig. 2.16. Just as in the case of derivation of formula (2.6), one can note that the sums of numbers in the pairs of shallow diagonals, which are parallel to the bottom left — top right diagonal, are multiples of 18. Therefore, in the 8×8 addition table the sum of multiples of three can be found as follows: $18 \cdot 2 + 18 \cdot 5 + 9 \cdot 8 = 18 \cdot (2 + 5) + 9 \cdot 8$. This sum is equal to 198 — the number confirmed computationally by the spreadsheet of Fig. 2.16 (cell G17).

Likewise, in the 11×11 addition table ($k = 4$, not pictured here) the sum of multiples of three can be found as follows:
$$24 \cdot 2 + 24 \cdot 5 + 24 \cdot 8 + 12 \cdot 11 = 24 \cdot (2 + 5 + 8) + 12 \cdot 11.$$

This sum is equal to 492 — the number can be confirmed computationally by the spreadsheet.

One more case may be considered to enable plausible generalization. In the 14×14 addition table ($k = 5$, not pictured here) the sum of multiples of three can be found as follows:
$$30 \cdot 2 + 30 \cdot 5 + 30 \cdot 8 + 30 \cdot 11 = 30 \cdot (2 + 5 + 8 + 11) + 12 \cdot 14.$$

This sum is equal to 1334 — the number can be confirmed computationally by the spreadsheet.

The above three special cases make it possible to generalize to the case $n = 3k - 1$ as follows:

$$\alpha(3k-1,3) = 6k[2+5+...+3(k-1)-1]+3k(3k-1)$$
$$= 6k(2+5+...+3k-4)+3k(3k-1)$$
$$= 6k \cdot \frac{2+(3k-4)}{2} \cdot (k-1)+3k(3k-1)$$
$$= 3k(3k^2-2k-3k+2+3k-1) = 3k(3k^2-2k+1).$$

That is,

$$\alpha(3k-1,3) = 3k(3k^2-2k+1). \tag{2.8}$$

In particular, $3k(3k^2-2k+1)\big|_{k=3} = 9 \cdot (27-6+1) = 198$ (Fig. 2.16, cell G17).

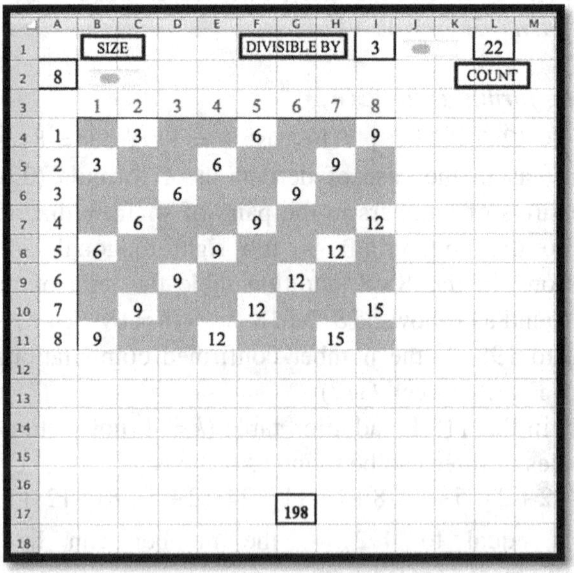

Fig. 2.16. Multiples of three in the 8×8 addition table.

Let $n = 3k = \{3, 6, 9, 12, ...\}$. When $k = 3$ (i.e., $n = 9$) this case is shown in Fig. 2.17. This time, special cases can be used differently. One can consider four cases: $k = 1$, $k = 2$, $k = 3$, and $k = 4$. The sums can be found computationally ($k = 3$ in Fig. 2.17, cell G17):

$$\alpha(3, 3) = 12, \ \alpha(6, 3) = 84, \ \alpha(9, 3) = 270, \ \alpha(12, 3) = 624.$$

	A	B	C	D	E	F	G	H	I	J	K	L	M
1		SIZE				DIVISIBLE BY		3				27	
2	9											COUNT	
3		1	2	3	4	5	6	7	8	9			
4	1		3			6				9			
5	2	3			6			9					
6	3			6			9			12			
7	4		6			9			12				
8	5	6			9			12					
9	6			9			12			15			
10	7		9			12			15				
11	8	9			12			15					
12	9			12			15			18			
13													
14													
15													
16													
17							270						
18													

Fig. 2.17. Multiples of three in the 9×9 addition table.

Assuming that $\alpha(3k, 3) = ak^3 + bk^2 + ck + d$, the coefficients a, b, c, and d can be found by using *Wolfram Alpha* as a tool for solving a system of four linear equations (Fig. 2.18). This yields the formula

$$\alpha(3k, 3) = 3k^2(3k + 1) \tag{2.9}$$

Finally, let $n = 3k + 1 = \{4, 7, 10, 13, ...\}$. When $k = 3$ (i.e., $n = 10$) this case is shown in Fig. 2.19. Once again, one can consider the following four cases: $k = 1$, $k = 2$, $k = 3$, and $k = 4$. The sums can be found computationally ($k = 3$ in Fig. 2.19, cell G17):

$$\alpha(4, 3) = 24, \ \alpha(7, 3) = 126, \ \alpha(10, 3) = 360, \ \alpha(13, 3) = 780.$$

Assuming that $\alpha(3k + 1, 3) = ak^3 + bk^2 + ck + d$, the coefficients a, b, c, and d can be found by using *Wolfram Alpha* as a tool for solving a system of four linear equations (Fig. 2.20). This yields the formula

$$\alpha(3k + 1, 3) = 3k(3k^2 + 4k + 1) \tag{2.10}$$

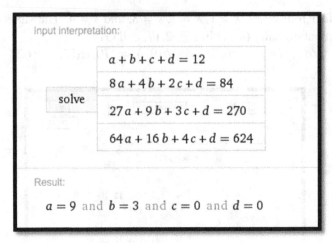

Input interpretation:

solve
$$a + b + c + d = 12$$
$$8a + 4b + 2c + d = 84$$
$$27a + 9b + 3c + d = 270$$
$$64a + 16b + 4c + d = 624$$

Result:

$$a = 9 \text{ and } b = 3 \text{ and } c = 0 \text{ and } d = 0$$

Fig. 2.18. Finding formula for $\alpha(3k, 3)$.

	A	B	C	D	E	F	G	H	I	J	K	L	M
1		SIZE				DIVISIBLE BY			3			33	
2	10											COUNT	
3		1	2	3	4	5	6	7	8	9	10		
4	1		3			6			9				
5	2	3			6			9			12		
6	3			6			9			12			
7	4		6			9			12				
8	5	6			9			12			15		
9	6			9			12			15			
10	7		9			12			15				
11	8	9			12			15			18		
12	9			12			15			18			
13	10		12			15			18				
14													
15													
16													
17						360							
18													

Fig. 2.19. Multiples of three in the 10×10 addition table.

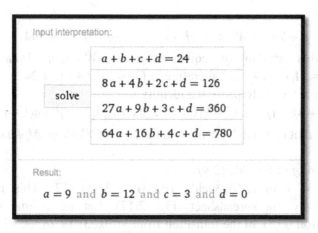

Fig. 2.20. Finding formula for $\alpha(3k+1,3)$.

2.7.2 *Proving formulas (2.8)-(2.10)*

As was shown in the case of $\alpha(n,2)$, the transition from n to $n + 1$ required knowledge of augmentation by the expression $f(n)$ such that $\alpha(n,2) + f(n) = \alpha(n+1,2)$. When $n = 2k$ we had

$$f(2k) = 12(k+1)^2 - (8k+6);$$

when $n = 2k + 1$ we had

$$f(2k+1) = 12k^2 + 4k + 2.$$

One can observe that in both cases, the sum of the multiples of two and its the augmentation in the transition from n to $n + 1$ is a quadratic function of n. This observation prompts a new method of proving formulas (2.8)-(2.10).

2.7.2.1 *Proof of formula (2.8)*

Setting $k = 1$ in (2.8) yields $\alpha(2,3) = 3(3-2+1) = 6$. This result is confirmed by the spreadsheet (Fig. 2.16). Let assume that the augmentation $g(k)$ in the transition from $\alpha(3k-1,3)$ to $\alpha(3(k+1)-1,3)$ has the form $g(k) = ak^2 + bk + c$. In order to find the coefficients a, b, and c, note that the numbers 6, 54, 198, and 492 represent computationally found sums of multiples of three in the addition tables of sizes 2, 5, 8, and 11, respectively. Therefore,

$$g(1) = 54 - 6 = 48, \; g(2) = 198 - 54 = 144, \; g(3) = 492 - 198 = 294.$$

That is,
$$a+b+c=48, 4a+2b+c=144, 9a+3b+c=294.$$

Solving this system of equations using *Wolfram Alpha* yields $a=27, b=51, c=24$, and, therefore, $g(k)=27k^2+51k+24$.

The final step is to prove the identity
$$3k(3k^2+4k+1)+27k^2+51k+24=3(k+1)[3(k+1)^2+4(k+1)+1],$$
something that can also be done with the help of *Wolfram Alpha*.

2.7.2.2 *Proof of formula (2.9)*

Setting $k=1$ in (2.9) yields $\alpha(3,3)=3(3+1)=12$. This result is confirmed by the spreadsheet (Fig. 2.17). Let us assume that the augmentation $g(k)$ in the transition from $\alpha(3k,3)$ to $\alpha(3(k+1),3)$ has the form $g(k)=ak^2+bk+c$. In order to find the coefficients a, b, and c, note that the numbers 12, 84, 270, and 624 represent computationally found sums of multiples of three in the addition tables of sizes 3, 6, 9, and 12, respectively. Therefore,
$$g(1)=84-12=72, g(2)=270-84=186, g(3)=624-270=354.$$

That is,
$$a+b+c=72, 4a+2b+c=186, 9a+3b+c=354.$$

Solving this system of equations using *Wolfram Alpha* yields $a=27, b=33, c=12$, and, therefore, $g(k)=27k^2+51k+24$.

The final step is to prove the identity
$$9k^3+3k^2+27k^2+33k+12=9(k+1)^3+3(k+1)^2,$$
something that can also be done with the help of *Wolfram Alpha*.

2.7.2.3 *Proof of formula (2.10)*

Setting $k=1$ in (2.10) yields $\alpha(4,3)=3(3+4+1)=24$. This result is confirmed by the spreadsheet (Fig. 2.17). Let assume that the augmentation $g(k)$ in the transition from $\alpha(3k+1,3)$ to $\alpha(3(k+1)+1,3)$ has the form $g(k)=ak^2+bk+c$. In order to find the coefficients a, b, and c, note that the numbers 24, 126, 360, and 780 represent computationally found sums of multiples of three in the addition tables of sizes 4, 7, 10, and 13, respectively. Therefore,
$$g(1)=126-24=102, g(2)=360-126=234, g(3)=780-360=420.$$

That is,
$$a + b + c = 102, \, 4a + 2b + c = 234, \, 9a + 3b + c = 420 \, .$$
Solving this system of equations using *Wolfram Alpha* yields $a = 27, b = 51, c = 24$, and, therefore, $g(k) = 27k^2 + 51k + 24$.

The final step is to prove the identity
$$3k(3k^2 + 4k + 1) + 27k^2 + 51k + 24 = 3(k+1)[3(k+1)^2 + 4(k+1) + 1] \, ,$$
something that can also be done with the help of *Wolfram Alpha*.

2.8 The Sum of the Multiples of Two in the *n*×*n* Multiplication Table

The next step could be to explore the sums of numbers in the multiplication table that display products with special divisibility properties. For example, to find the sum of all even numbers in the 10×10 multiplication table and then, by using tables of different size, to try to develop formulas for the sums related to other tables. As in the case of addition tables, this kind of explorations paves a way into algebra and the appropriate use of technology in carrying out symbolic computations through a numerical approach to such mundane concepts as the addition and multiplication tables.

Apparently, all even numbers in the multiplication tables can be arranged in two groups: one group consists of those numbers that belong to complete rows of products, another group consist of those numbers that belong to incomplete rows of products. Because of that, all *n*×*n* multiplication tables can be divided in two groups depending on whether *n* is even ($n = 2k$) or odd ($n = 2k - 1$).

2.8.1 *The case* $n = 2k$

Let $k = 5$. As shown in Fig. 2.21, $\sigma(10, 2)$ can be found by adding two groups of sums: one group with five addends and another group with ten addends. The elements of both groups reside in five lines. Let us denote $\sigma_5(10, 2)$ and $\sigma_{10}(10, 2)$ the sums of the elements of the first and the second groups, respectively. In the first group (with five addends), each term is an odd multiple of the sum
$$2 + 4 + 6 + 8 + 10 = 2(1 + 2 + 3 + 4 + 5) = 2 \cdot \frac{5 \cdot 6}{2} = 5 \cdot 6 \, .$$

Therefore,

$$\sigma_5(10, 2) = 1 \cdot (5 \cdot 6) + 3 \cdot (5 \cdot 6) + 5 \cdot (5 \cdot 6) + 7 \cdot (5 \cdot 6) + 9 \cdot (5 \cdot 6)$$
$$= (5 \cdot 6) \cdot (1 + 3 + 5 + 7 + 9) = (5 \cdot 6) \cdot 5^2 = 5^3 \cdot (5 + 1).$$

	A	B	C	D	E	F	G	H	I	J	K	L	M
1		SIZE				DIVISIBLE BY			2			75	
2	10											COUNT	
3		1	2	3	4	5	6	7	8	9	10		
4	1		2		4		6		8		10		
5	2	2	4	6	8	10	12	14	16	18	20		
6	3		6		12		18		24		30		
7	4	4	8	12	16	20	24	28	32	36	40		
8	5		10		20		30		40		50		
9	6	6	12	18	24	30	36	42	48	54	60		
10	7		14		28		42		56		70		
11	8	8	16	24	32	40	48	56	64	72	80		
12	9		18		36		54		72		90		
13	10	10	20	30	40	50	60	70	80	90	100		
14													
15													
16													
17						2400							
18													

Fig. 2.21. Multiples of two in the multiplication table of even size.

In the second group (with ten addends), each term is a multiple of the sum

$$2 + 4 + 6 + \ldots + 20 = 2(1 + 2 + 3 + \ldots + 10) = 2 \cdot \frac{10 \cdot 11}{2} = 10 \cdot 11.$$

Therefore,

$$\sigma_{10}(10, 2) = 1 \cdot (10 \cdot 11) + 2(5 \cdot 6) + 3(10 \cdot 11) + 4(10 \cdot 11) + 5(10 \cdot 11) =$$
$$10 \cdot 11 \cdot (1 + 2 + 3 + 4 + 5) = 10 \cdot 11 \cdot \frac{5 \cdot 6}{2} = 5^2(5 + 1)(2 \cdot 5 + 1).$$

Adding the terms of both groups yields

$$\sigma(10, 2) = \sigma_5(10, 2) + \sigma_{10}(10, 2) = 5^3(5 + 1) + 5^2(5 + 1)(2 \cdot 5 + 1)$$
$$= 5^2(5 + 1)(5 + 2 \cdot 5 + 1) = 5^2(5 + 1)(3 \cdot 5 + 1).$$

That is, when $k = 5$ we have

$$\sigma(2 \cdot 5, 2) = 5^2(5 + 1)(3 \cdot 5 + 1).$$

The last relation allows for an obvious generalization

$$\sigma(2k,2) = k^2(k+1)(3k+1). \qquad (2.11)$$

In particular, $\sigma(10, 2) = 25 \cdot 6 \cdot 16 = 2400$ — the result confirmed by the spreadsheet (Fig. 2.21, cell G17). However, empirically found formula (2.11) has yet to be proved.

2.8.2 *The case n = 2k − 1*

Continuing in the same vein, the sum $\sigma(9, 2)$, as shown in Fig. 2.22, can be found by adding two groups of sums: one group with four addends and another group with nine addends. In the first group there are five terms; in the second group there are four terms. The elements of the first group reside in five lines each of which has four terms. The elements of the second group reside in four lines, each of which has nine terms. Let's denote the sums of the elements in the former and the latter groups $\sigma_4(9, 2)$ and $\sigma_9(9, 2)$, respectively.

The first sum included in $\sigma_4(9, 2)$ is

$$2 + 4 + 6 + 8 = 2(1 + 2 + 3 + 4) = 2 \cdot \frac{4 \cdot 5}{2} = 4 \cdot 5.$$

Therefore,

$$\sigma_4(9, 2) = 1 \cdot (4 \cdot 5) + 3 \cdot (4 \cdot 5) + 5 \cdot (4 \cdot 5) + 7 \cdot (4 \cdot 5) + 9 \cdot (4 \cdot 5)$$

$$= 4 \cdot 5 \cdot (1 + 3 + 5 + 7 + 9) = 4 \cdot 5 \cdot 5^2 = 5^3(5 - 1).$$

The first sum included in $\sigma_9(9, 2)$ is

$$2 + 4 + 6 + ... + 18 = 2(1 + 2 + 3 + ... + 9) = 2 \cdot \frac{9 \cdot 10}{2} = 9 \cdot 10.$$

Therefore,

$$\sigma_9(9, 2) = 1 \cdot (9 \cdot 10) + 2 \cdot (9 \cdot 10) + 3 \cdot (9 \cdot 10) + 4 \cdot (9 \cdot 10)$$

$$= 9 \cdot 10 \cdot (1 + 2 + 3 + 4) = 9 \cdot 10 \cdot \frac{4 \cdot 5}{2} = 9 \cdot 5 \cdot 4 \cdot 5 = 5^2 \cdot (5 - 1)(2 \cdot 5 - 1).$$

Adding $\sigma_4(9, 2)$ and $\sigma_9(9, 2)$ yields

$$\sigma(9, 2) = \sigma_4(9, 2) + \sigma_9(9, 2) = 5^3 \cdot (5 - 1) + 5^2 \cdot (5 - 1) \cdot (2 \cdot 5 - 1)$$

$$= 5^2 \cdot (5 - 1) \cdot (3 \cdot 5 - 1).$$

That is, when $k = 5$ we have $\sigma(2 \cdot 5 - 1, 2) = 5^2 \cdot (5 - 1) \cdot (3 \cdot 5 - 1)$. Generalizing from the last relation yields

$$\sigma(2k-1,2) = k^2(k-1)(3k-1). \qquad (2.12)$$

In particular, $\sigma(9,2) = 5^2 \cdot 4 \cdot 14 = 1400$ — the result is confirmed computationally by the spreadsheet (Fig. 2.22, cell G17). However, empirically found formula (2.12) has yet to be proved.

	A	B	C	D	E	F	G	H	I	J	K	L	M
1		SIZE				DIVISIBLE BY		2				56	
2	9											COUNT	
3		1	2	3	4	5	6	7	8	9			
4	1		2		4		6		8				
5	2	2	4	6	8	10	12	14	16	18			
6	3		6		12		18		24				
7	4	4	8	12	16	20	24	28	32	36			
8	5		10		20		30		40				
9	6	6	12	18	24	30	36	42	48	54			
10	7		14		28		42		56				
11	8	8	16	24	32	40	48	56	64	72			
12	9		18		36		54		72				
13													
14													
15													
16													
17							1400						
18													

Fig. 2.22. Multiples of two in the multiplication table of odd size.

2.8.3 Proof of formula (2.11)

Formula (2.11) can be proved by the method of mathematical induction in much the same way as the formulas associated with addition tables. When $k = 1$ we have $\sigma(2,2) = 1 \cdot (1+1)(3 \cdot 1 + 1) = 8$. This result is confirmed by the spreadsheet of Fig. 2.21 (range B4:C5) — there are three even numbers in the 2×2 multiplication table; namely, 2, 2, and 4. Alternatively, the three numbers represent the gnomon of rank two of the multiplication table, with the sum $2^3 = 8$.

Transition from k to $k + 1$ requires the augmentation to be expressed in terms of k. To this end, let $k = 4$. As shown in Fig. 2.21, the transition from 8×8 multiplication table to 10×10 multiplication table brings about the sum

$$2 \cdot (18 + 36 + 54 + 72) = 2 \cdot 18 \cdot (1 + 2 + 3 + 4)$$

$$= 2 \cdot 18 \cdot \frac{4 \cdot 5}{2} = 2 \cdot 9 \cdot 4 \cdot 5 = 2 \cdot (2 \cdot 4 + 1) \cdot 4 \cdot (4 + 1)$$

and the full gnomon of rank 10 which is equal to $10^3 = [2 \cdot (4 + 1)]^3$. That is, transition from $k = 4$ to $k = 5$ adds to the table the sum $2 \cdot (2 \cdot 4 + 1) \cdot 4 \cdot (4 + 1) + 8 \cdot (4 + 1)^3$. Therefore, one has to prove that $\sigma(2k, k) + 2k(k + 1)(2k + 1) + 8(k + 1)^3 = \sigma(2k + 2, k)$. Put another way, one has to prove the identity

$$k^2(k + 1)(3k + 1) + 2k(k + 1)(2k + 1) + 8(k + 1)^3 = (k + 1)^2(k + 2)(3k + 4).$$

As before, the last identity can be confirmed by *Wolfram Alpha*.

2.8.4 *Proof of formula (2.12)*

Setting $k = 1$ in (2.12) yields zero; indeed, the multiplication table of size one does not have an even product. Let $k = 4$. As shown in Fig. 2.22, the transition from 7×7 table to 9×9 table brings about the following sum: $2 \cdot 4 \cdot (4 + 1) \cdot (2 \cdot 4 + 1) + (2 \cdot 4)^3$. That is, the augmentation in terms of k is the sum $2k(k + 1)(2k + 1) + 8k^3$. It remains to be proved that

$$\sigma(2k - 1, 2) + 2k(k + 1)(2k + 1) + 8k^3 = \sigma(2k + 1, 2).$$

Put another way, one has to prove the identity

$$k^2(k - 1)(3k - 1) + 8k^3 + 2k(k + 1)(2k + 1) = (k + 1)^2 k(3k + 2).$$

Wolfram Alpha confirms this identity and this completes proving formula (2.12).

Remark 2.13. Formulas (2.11) and (2.12) can be found differently by noting that the sum of odd numbers both in $2k \times 2k$ and $(2k - 1) \times (2k - 1)$ multiplication tables equals k^4. As shown in Fig. 2.23, a transition from an odd size multiplication table to an even size does not add any odd product, as all numbers within the gnomon of rank $2k$ are multiples of $2k$. In Fig. 2.23, the sum of odd products consists of five sums each of which is an odd multiple of the sum in the first line. So, denoting $\sigma(n, odd)$ the sum of odd numbers in the $n \times n$ multiplication table, we have both for $n = 2k$ and $n = 2k - 1$

$$\sigma(n, odd) = 1 \cdot (1 + 3 + 5 + ... + 2k - 1) + 3 \cdot (1 + 3 + 5 + ... + 2k - 1) + ...$$

$$+ (2k - 1) \cdot (1 + 3 + 5 + ... + 2k - 1) = (1 + 3 + 5 + ... + 2k - 1)^2 = k^4.$$

Therefore,

$$\sigma(2k, 2) = \sigma(2k) - \sigma(2k, odd) = [\frac{2k(2k+1)}{2}]^2 - k^4$$

$$= k^2(2k+1)^2 - k^4 = k^2[(2k+1)^2 - k^2] = k^2(k+1)(3k+1)$$

and

$$\sigma(2k-1, 2) = \sigma(2k-1) - \sigma(2k-1, odd) = [\frac{(2k-1)\cdot 2k}{2}]^2 - k^4$$

$$= (2k-1)^2 k^2 - k^4 = k^2[(2k-1)^2 - k^2] = k^2(k-1)(3k-1).$$

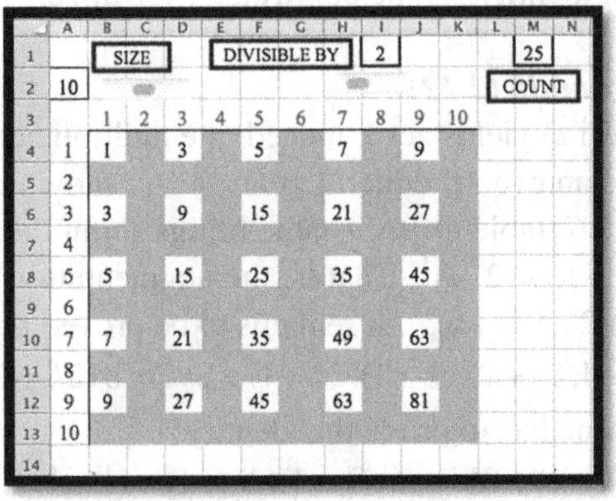

Fig. 2.23. Odd products in the multiplication table of size ten.

2.9   The Sum of the Multiples of Three in the $n \times n$ Multiplication Table

In order to find $\sigma(n, 3)$, three cases need to be considered: $n = 3k$, $n = 3k + 1$, and $n = 3k - 1$. These three cases will be considered separately. When $k = 3$, the three special cases for are presented in Figs. 2.24, 2.25, and 2.26, respectively. In each case, the spreadsheet automatically calculates the corresponding sum.

2.9.1   *The case n = 3k*

Let $k = 3$. As shown in Fig. 2.24, all products divisible by three reside on three horizontal and three vertical lines the intersections of which comprise common products. When adding numbers that belong to the six lines, one has to subtract those residing at the intersections. In that way,

$$\sigma(9,3) = 2 \cdot [1 \cdot (3+6+9+...+27) + 2 \cdot (3+6+9+...+27)$$

$$+3 \cdot (3+6+9+...+27)] - 1 \cdot (9+18+27) - 2 \cdot (9+18+27)$$

$$-3 \cdot (9+18+27) = 2 \cdot (3+6+9+...+27) \cdot (1+2+3)$$

$$-(9+18+27) \cdot (1+2+3) = 6 \cdot (1+2+3+...+9) \cdot (1+2+3)$$

$$-9 \cdot (1+2+3)^2 = 6 \cdot \frac{9 \cdot 10}{2} \cdot \frac{3 \cdot 4}{2} - 9 \cdot \left(\frac{3 \cdot 4}{2}\right)^2 = 1296.$$

	A	B	C	D	E	F	G	H	I	J	K	L	M
1			SIZE				DIVISIBLE BY		3			45	
2	9											COUNT	
3		1	2	3	4	5	6	7	8	9			
4	1			3			6			9			
5	2			6			12			18			
6	3	3	6	9	12	15	18	21	24	27			
7	4			12			24			36			
8	5			15			30			45			
9	6	6	12	18	24	30	36	42	48	54			
10	7			21			42			63			
11	8			24			48			72			
12	9	9	18	27	36	45	54	63	72	81			
13													
14													
15													
16													
17							1296						
18													

Fig. 2.24. Multiples of three in the $3 \cdot 3 \times 3 \cdot 3$ -multiplication table.

Generalizing to the case $n = 3k$ yields

$$\sigma(3k,3) = 2 \cdot 3 \cdot (1+2+...+3k) \cdot (1+2+...+k) - 9 \cdot (1+2+...+k)^2$$

$$= 3 \cdot (1+2+...+k)[2 \cdot (1+2+...+3k) - 3 \cdot (1+2+...+k)]$$

$$= 3 \cdot \frac{k(k+1)}{2} \cdot [2 \cdot \frac{3k(3k+1)}{2} - 3 \cdot \frac{k(k+1)}{2}] = \frac{9}{4}k^2(k+1)(5k+1).$$

That is,

$$\sigma(3k,3) = \frac{9}{4}k^2(k+1)(5k+1). \tag{2.13}$$

In particular, $\sigma(9,3) = \frac{9}{4} \cdot 3^2 \cdot 4 \cdot 16 = 1296$. This result is confirmed computationally by using the spreadsheet (Fig. 2.24, cell G17).

2.9.2 *The case n = 3k + 1*

As shown in Fig. 2.25, just as in the case $n = 3k$, all products divisible by three reside on three horizontal and three vertical lines the intersections of which comprise common products. Once again, when adding numbers that belong to the six lines, one has to subtract those residing at the intersections. In that way,

$$2 \cdot [1 \cdot (3 + 6 + 9 + \ldots + 30) + 2 \cdot (3 + 6 + 9 + \ldots + 30) + 3 \cdot (3 + 6 + 9 + \ldots + 30)]$$
$$-1 \cdot (9 + 18 + 27) - 2 \cdot (9 + 18 + 27) - 3 \cdot (9 + 18 + 27)$$
$$= 2 \cdot (3 + 6 + 9 + \ldots + 30) \cdot (1 + 2 + 3) - (9 + 18 + 27) \cdot (1 + 2 + 3)$$
$$= 6 \cdot (1 + 2 + 3 + \ldots + 10) \cdot (1 + 2 + 3) - 9 \cdot (1 + 2 + 3)^2$$
$$= 6 \cdot \frac{10 \cdot 11}{2} \cdot \frac{3 \cdot 4}{2} - 9 \cdot \left(\frac{3 \cdot 4}{2}\right)^2 = 1656.$$

Generalizing to the case $n = 3k + 1$ yields
$$\sigma(3k + 1, 3) = 2 \cdot 3 \cdot [1 + 2 + \ldots + (3k + 1)] \cdot (1 + 2 + \ldots + k) - 9 \cdot (1 + 2 + \ldots + k)^2$$
$$= 3 \cdot (1 + 2 + \ldots + k)[2 \cdot (1 + 2 + \ldots + 3k + 1) - 3 \cdot (1 + 2 + \ldots + k)]$$
$$= 3 \cdot \frac{k(k+1)}{2} \cdot [2 \cdot \frac{(3k+1)(3k+2)}{2} - 3 \cdot \frac{k(k+1)}{2}] = \frac{3}{4} k(k+1)(15k^2 + 15k + 4).$$

That is,

$$\sigma(3k + 1, 3) = \frac{3}{4} k(k+1)(15k^2 + 15k + 4). \qquad (2.14)$$

In particular, $\sigma(10, 3) = \frac{3}{4} \cdot 3 \cdot 4 \cdot (15 \cdot 9 + 15 \cdot 3 + 4) = 1656$. This result is confirmed computationally by using the spreadsheet (Fig. 2.25, cell G17).

	SIZE				DIVISIBLE BY		3			51	
10										COUNT	
	1	2	3	4	5	6	7	8	9	10	
1			3			6			9		
2			6			12			18		
3	3	6	9	12	15	18	21	24	27	30	
4			12			24			36		
5			15			30			45		
6	6	12	18	24	30	36	42	48	54	60	
7			21			42			63		
8			24			48			72		
9	9	18	27	36	45	54	63	72	81	90	
10			30			60			90		
					1656						

Fig. 2.25. Multiples of three in the $(3 \cdot 3 + 1) \times (3 \cdot 3 + 1)$-multiplication table.

	SIZE				DIVISIBLE BY		3			28	
8										COUNT	
	1	2	3	4	5	6	7	8			
1			3			6					
2			6			12					
3	3	6	9	12	15	18	21	24			
4			12			24					
5			15			30					
6	6	12	18	24	30	36	42	48			
7			21			42					
8			24			48					
					567						

Fig. 2.26. Multiples of three in the $(3 \cdot 3 - 1) \times (3 \cdot 3 - 1)$-multiplication table.

2.9.3 The case n = 3k − 1

As shown in Fig. 2.26, all products divisible by three reside on two horizontal and two vertical lines the intersections of which comprise common products. When adding numbers that belong to the four lines, one has to subtract those residing at the intersections. In that way,

$$2 \cdot (3 + 6 + \ldots + 24) \cdot (1 + 2) - (9 + 18) \cdot (1 + 2)$$

$$= 6 \cdot (1 + 2 + \ldots + 8) \cdot (1 + 2) - 9 \cdot (1 + 2)^2 = 18 \cdot \frac{8 \cdot 9}{2} - 81 = 567 \, .$$

Generalizing to the case $n = 3k - 1$ yields

$$\sigma(3k - 1, 3) = 2 \cdot 3 \cdot [1 + 2 + \ldots + (3k - 1)][(1 + 2 + 3 + \ldots + (k - 1)]$$

$$-9 \cdot [1 + 2 + \ldots + (k - 1)]^2 = 6 \frac{(3k - 1) \cdot 3k}{2} \cdot \frac{(k - 1) \cdot k}{2}$$

$$-9 \cdot \frac{(k - 1)^2 \cdot k^2}{2} = \frac{9}{4} \cdot k^2 (k - 1)(5k - 1) \, .$$

That is,

$$\sigma(3k - 1, 3) = \frac{9}{4} k^2 (k - 1)(5k - 1) \, . \tag{2.15}$$

In particular, $\sigma(8, 3) = \frac{9}{4} \cdot 9 \cdot 2 \cdot 14 = 567$. This result is confirmed computationally using the spreadsheet (Fig. 2.26, cell G17).

2.9.4 *Proof of formula (2.13)*

Formula (2.13) can also be proved by the method of mathematical induction. To this end, setting $k = 1$ in (2.13) yields $\sigma(3, 3) = \frac{9}{4} \cdot 12 = 27 = 3^3$ — the gnomon of rank three in the multiplication table. In order to show the transition from k to $k + 1$, one has to describe the augmentation of the multiples of three in the table as follows:

$$\sigma[3(k + 1), 3] = \sigma(3k, 3) + 2 \cdot 3(k + 1)[1 + 2 + \ldots + 3(k + 1)] - 9(k + 1)^2$$

$$+ 3 \cdot 2(1 + 2 + \ldots + k)(3k + 1 + 3k + 2) = \sigma(3k, 3) + 6(k + 1) \frac{(3k + 3)(3k + 4)}{2}$$

$$-9(k + 1)^2 + 6 \frac{k(k + 1)}{2}(6k + 3) = \sigma(3k, 3) + 9(k + 1)^2(3k + 3) + 9(k + 1)^2$$

$$-9(k + 1)^2 + 9k(k + 1)(2k + 1) = \sigma(3k, 3) = 9(k + 1)(5k^2 + 7k + 3) \, .$$

Now, assuming that formula (2.13) is true, one has to prove the identity

$$\frac{9}{4} \cdot k^2(k+1)(5k+1) + 9(k+1)(5k^2+7k+3) = \frac{9}{4} \cdot (k+1)^2(k+1)(5k+6)$$

This identity can be confirmed by using *Wolfram Alpha*. Formula (2.13) has been proved. Similarly, formulas (2.15) and (2.16) can be proved.

Chapter 3

Spreadsheet as a 3-D Modeling Tool

3.1 Introduction

In the first two chapters, many mathematical activities were based on a spreadsheet facility to numerically model one- and two-dimensional problems such as generating arithmetic sequences of the given length (Chapter 1) or recognizing numbers with special properties within the addition/multiplication table (Chapter 2). In this chapter, a spreadsheet will be used for modeling three-dimensional problems that can be found in (or relevant to) the pre-college mathematics curriculum. These problems are described by linear Diophantine equations in three unknowns. One of the goals of this chapter is to show how teacher candidates can appreciate the idea that spreadsheets belong to "relevant external mathematical resources ... [that can be used] to *pose* or solve problems" [Common Core State Standards, 2010, p. 7, italics added]. To this end, modeling data provided by a spreadsheet will be used to pose new problems. The focus will be on posing the so-called technology-immune/technology-enabled (TITE) problems [Abramovich, 2014], the process of solving of which is both dependent on the use of a computer and resistant to symbolic computations as a workable method of finding an answer. The importance of TITE problems for the modern day school mathematics curriculum is two-fold: they support the worldwide emphasis on using technology for mathematics teaching and learning and take into account the availability of powerful computer programs, like *Wolfram Alpha*, capable of, in response to a natural language input, easily solving traditional problems by performing symbolic computations of different levels of complexity. For example, the proof of formula (2.5) in Chapter 2 may be considered a TITE problem as technology cannot carry out the whole proof at the push of a button; yet, using *Wolfram Alpha* was essential in completing the inductive transfer. In that way, in the digital era, the idea of TITE problems supports the notion that "technology should be used as one of a variety of approaches and tools for creating mathematical understanding" [Western and Northern Canadian Protocol, 2008, p. 9].

3.2   Spreadsheet Modeling of a Diophantine Equation with Three Unknowns

To begin, consider the following modification of a problem once recommended for grade 8 [New York State Education Department, 1998, p. 114].

Problem 3.1. *Three segments are given whose lengths are 2, 3, and 5 centimeters. Using any of the given lengths as many times as you wish, determine how many segments of length 16 centimeters can be constructed.*

The following linear Diophantine equation

$$2x + 3y + 5z = 16 \qquad (3.1)$$

represents a (three-dimensional) mathematical model of the situation described in this problem where x, y, and z are unknown non-negative integers. Equation (3.1) can be interpreted as the partition of number 16 into the summands two, three, and five. Below the notation $D(n; a, b, c)$ will be used to express the number of ways the number n can be partitioned into the summands a, b, and c. It is called [Comtet, 1974] the denumerant of n with respect to the sequence a, b, and c. The denumerant is an example of a verbally defined function. When the number of summands is not greater than three, there exist closed formulas for a denumerant [Comtet, 1974]. In this chapter a spreadsheet will be used to compute denumerants. Therefore, the task posed by Problem 3.1 is to find $D(16; 2, 3, 5)$.

For example, one can describe the number of whole number solutions to the equation $2x + 3y + 5z = 2$ as $D(2; 2, 3, 5) = 1$ meaning that the equation has a single solution, namely, $x = 1$, $y = 0$, $z = 0$. Likewise, $D(5; 2, 3, 5) = 2$ meaning that there exists two solutions to the equation $2x + 3y + 5z = 5$, namely, $x = 1$, $y = 1$, $z = 0$, and $x = 0$, $y = 0$, $z = 1$. Solutions to Eq. (3.1), alternatively, the value of $D(16; 2, 3, 5)$, can be found by using a spreadsheet as a three-dimensional modeling tool (Fig. 3.1).

To this end, Eq. (3.1) can be rewritten in the form $z = [16 - (2x + 3y)] / 5$ from where the following condition (to ensure that z is a non-negative integer) follows: the difference $16 - (2x + 3y)$ is a non-negative multiple of five. This condition can be expressed through the spreadsheet formula

=IF(OR(D$3=" ", $C4=" "), " ",IF(AND($A$1-
(D$3*$E$1+$C4*$F$1)>=0,MOD($A$1-
(D$3*$E$1+$C4*$F$1),$G$1)=0), ($A$1-
(D$3*$E$1+$C4*$F$1))/$G$1," "))

defined in cell D4 (Fig. 3.1) and replicated across the whole table (to cell L8) to generate the values of z satisfying Eq. (3.1). Its first part including the logical function OR leaves cells blank when the values of the variables x and y defined, respectively, in row 3 (beginning from cell D3) and column C (beginning from cell C4) are beyond their range. To clarify, note that the maximum number of twos to make 16 is eight (the largest number in row 3) and the maximum number of threes to make 16 is five (the largest number in column C). To this end, after cells D3 and C4 are entered with zeros, in cells E3 and C5 the following formulas are defined and replicated through row 3 and column C, respectively, =IF(D3<INT($A1/$E1),D3+1," ") and =IF(C4<INT(A$1/F$1),C4+1," ").

	A	B	C	D	E	F	G	H	I	J	K	L	M
1	16		side		2	3	5		segments				
2													
3				0	1	2	3	4	5	6	7	8	
4			0				2					0	
5			1					1					
6			2	2				0					
7			3	1									
8			4		0								
9			5										
10													

Fig. 3.1. Modeling Eq. (3.1).

Cell A1 is entered with the number 16; cells E1, F1, and G1 are entered with the coefficients of Eq. (3.1). For example, cells H3, C5, and H5 show the triple (4, 1, 1) — one of the solutions to Problem 3.1. This solution means $16 = 4 \cdot 2 + 1 \cdot 3 + 1 \cdot 5$; that is, a 16-cm long segment can be made out of four 2-cm segments, one 3-cm segment, and one 5-cm segment. The total number of ways of making a 16-cm segment is seven — the number of non-empty cells in the range D4:L8 (Fig. 3.1). By interpreting numeric information associated with this range, one can list all the solutions as follows:

$$16 = 3 \cdot 2 + 0 \cdot 3 + 2 \cdot 5; 16 = 8 \cdot 2 + 0 \cdot 3 + 0 \cdot 5; 16 = 4 \cdot 2 + 1 \cdot 3 + 1 \cdot 5;$$

$$16 = 0 \cdot 2 + 2 \cdot 3 + 2 \cdot 5; \ 16 = 1 \cdot 2 + 3 \cdot 3 + 1 \cdot 5; 16 = 2 \cdot 2 + 4 \cdot 3 + 0 \cdot 5;$$

$$16 = 8 \cdot 2 + 0 \cdot 3 + 0 \cdot 5.$$

Remark 3.1. Using *Wolfram Alpha* one can verify the correctness of spreadsheet modeling. To this end, one can enter the command "solve $2x + 3y + 5z = 16$, x>=0, y >=0, z>=0, over the integers" into the input box of the program (Fig. 3.2) to enable the production of the same seven solutions presented as a union of triples (x, y, z) satisfying Eq. (3.1). Such use of *Wolfram Alpha* plays the dual rule: it verifies the correctness of spreadsheet modeling of Eq. (3.1) and presents all seven solutions in an alternative, more lucid way which does not expect learners to interpret numerical information generated by the spreadsheet.

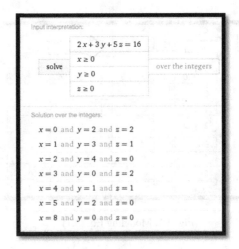

Fig. 3.2. *Wolfram Alpha* as an alternative to a spreadsheet.

Remark 3.2. Note that analytic formulas for calculating denumerants are based on the function $\|x\|$ — the integer closest to x — and they are not unique even for a given set of summands. For example [Comtet, 1974, p. 110],

$$D(n;1,2,3) = \left\| (n+3)^2 /12 \right\| = \left\| (n+2)(n+4)/12 \right\|$$

$$= \left\| (n^2 + 6n + 7)/12 \right\| = \left\| (n^2 + 6n + 10)/12 \right\|.$$

Each of the formulas yields $D(5;1,2,3) = 5$ — the result being in agreement with the fact that the equation $x + 2y + 3z = 5$ has exactly five solutions: (0, 1, 1), (2, 0, 1), (3, 1, 0), (1, 2, 0), and (5, 0, 0). These solutions are not difficult to obtain by trial and error. Likewise, the use of the spreadsheet pictured in Fig. 3.1 can confirm the same result computationally.

Analyzing the above four closed formulas for calculating the denumerant $D(n; 1,2,3)$, one may note that another formula, $D(n;1,2,3) = \left\| (n^2 + 6n + 5)/12 \right\|$, works for $n = 5$. Indeed, we have $\left\| (5^2 + 6 \cdot 5 + 5)/12 \right\| = 5$. However, when $n = 12$ this new formula yields $\left\| 221/12 \right\| = 18$. At the same time, the spreadsheet of Fig. 3.1 yields 19 whole number solutions to the equation $x + 2y + 3z = 12$ and this result is in agreement with each of the above four formulas; for example, $D(12;1,2,3) = \left\| (12+3)^2 /12 \right\| = \left\| 225/12 \right\| = \left\| 18.75 \right\| = 19$. Therefore, the fifth formula conjectured towards a seemingly easy augmentation of the above four formulas shows a real sophistication of purely mathematical propositions. Unlike the case $n = 5$, the case $n = 12$ turned out to be sufficient not to accept the formula as a valid hypothesis and it points out how one can "explain whether something is valid by proving or invalid by indicating a counterexample" [Takahashi *et al.*, 2006, p. 27]. Nonetheless, the confirmation of the validity of the formulas through numerical evidence may not be considered a proof in the strict mathematical sense.

3.3 3-D Spreadsheet Modeling Involving Parameters

Apparently, there is nothing special about the lengths of the segments involved in Problem 3.1 and, therefore, the segments may become variable entities. In this case, the following Diophantine equation

$$ax + by + cz = n \tag{3.2}$$

represents a generalized model that describes various three-dimensional problems; in particular, a problem with three contributing segments. As an example, consider the following extension of Problem 3.1.

Problem 3.2. *Three segments are given whose lengths are 2, 3, and 4 centimeters. Using any of the given lengths as many times as you wish, determine how many segments of length not greater than 12 centimeters can be constructed.*

In this problem, the length of the combined segment will be a variable defined in cell A1, the content of which can be controlled by a slider. In addition, another method of programming a spreadsheet will be suggested to allow for the variation of two parameters of the problem. A spreadsheet that generates all integral solutions to Eq. (3.2) (with a possibility to alter coefficients a, b, c, and the value of its right-hand side n by using sliders) is pictured in Fig. 3.3. It generates two sets of data simultaneously — solutions (x, y, z) to Eq. (3.2) and the corresponding values of $D(n; 2, 3, 4)$ in the range $2 \leq n \leq 12$. The triple of the contributing segment lengths $(a, b, c) = (2, 3, 4)$ and the length n of the combined segment are displayed in cells E1, F1, G1, and A1, respectively (in Fig. 3.3, the content of cell A1 is 12 — the largest possible in the context of Problem 3. 2). The spreadsheet formula

=IF(D$3*$E$1+$C4*$F$1+$C$3*$G$1=$A$1,
$C$3,IF($C$3>0,D4," ")),

defined in cell D4 and replicated to cell L12, enables one to solve Eq. (3.2) by generating appropriate values of one of the unknowns (namely, z) through a reference to cell C3 where the value of the variable z varies over its admissible range. The ranges for the unknowns x and y are, respectively, [D3:L3] and [C4:C12]. The above spreadsheet formula includes a circular reference; this makes it possible to store in the range [D4:L12] all appropriate values of the variable z related to the value of cell A1 as z varies in cell C3. That is, given n (cell A1), one generates all solutions to Eq. (3.2) by varying the value of z through a slider that controls cell C3. For example, when the content of cell A1 is 11, one has to alter the content of cell C3 in the range [0, 2]. That is, the 4-cm segment may be used in the construction of an 11-cm segment either twice, or once, or not at all. As the length 11 changes to that of 12, the admissible range for z becomes [0, 3].

More specifically, when the content of cell A1 is 12, the first value of cell C3 is zero (not shown in Fig. 3.3) — the spreadsheet interactively generates three zeros in the range [D4:L12] (shown in Fig. 3.3). As the value of cell C3 changes to one, the circular reference in the last spreadsheet formula keeps three solutions with zero values of z and adds two new solutions with $z = 1$. Likewise, when the value of cell C3 is two, the spreadsheet preserves five previous solutions (for $z = 0, 1$) and generates one more solution with $z = 2$. Finally, when the value of cell C3 is three (shown in Fig. 3.3) we have the last solution with $z = 3$ (i.e., $x = y = 0$, $z = 3; 12 = 3 \cdot 4$). In all, we have seven solutions to the equation $2x + 3y + 4z = 12$. As will be explained in the next paragraph, this number of solutions is recorded in cell N12.

	A	B	C	D	E	F	G	H	I	J	K	L	M	N
1	12		side		2	3	4		segments			33		
2													2	1
3			3	0	1	2	3	4	5	6	7	8	3	1
4			0	3		2		1		0			4	2
5			1										5	1
6			2		1		0						6	3
7			3										7	2
8			4	0									8	4
9			5										9	3
10			6										10	5
11			7										11	4
12			8										12	7

Fig. 3.3. Spreadsheet as a tool for solving Problem 3.2.

Just as the spreadsheet of Fig. 3.1, the spreadsheet pictured in Fig. 3.3 displays seven solutions to Eq. (3.2) with $n = 12$, $a = 2$, $b = 3$, $c = 4$. As the value of n changes, the number of solutions for each n is recorded in column U. In other words, the range [M2:N12] serves as the table representation of the denumerant function $D(n; 2, 3, 4)$ — column T contains values of n (all that the variable content of cell A1 assumes) while column U contains numerical values of the function. These values are generated through the spreadsheet formula

=IF(A\$1=1," ",IF(A\$1=M2,COUNT(D\$4:L\$12),N2))

defined in cell N2 and replicated to cell N12. Once again, the use of a circular reference in this formula (that is, a relative reference to cell N2) enables the spreadsheet to keep in column N the values of the already computed $D(i; 2, 3, 4)$, $i = 2, 3, ..., n$, unchanged as the content of cell A1 changes from $i = n$ to $i = n+1$. One can see (Fig. 3.3) that the sum of numbers in the range [N2:N12] is equal to 33 (cell L1) — the number of segments of length not greater than 12 that can be constructed out of the segments 2, 3, and 4.

Figure 3.4 shows the graph of the function $D(n; 2, 3, 4)$ constructed by the spreadsheet in the range $2 \leq n \leq 12$. This graph reflects many interesting quantitative situations associated with Problem 3.2 including a non-monotonous growth of the denumerant function as n increases. In particular, this example shows how at the high school level the study of functions "to describe how one quantity of interest depends on another" [Common Core State Standards, 2010, p. 7] can be mediated both by geometric (or any other) context and technology.

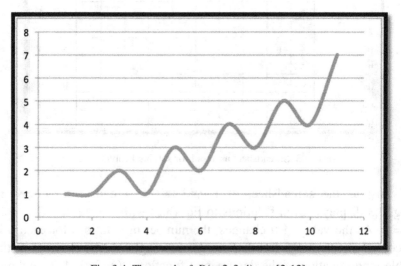

Fig. 3.4. The graph of $D(n; 2, 3, 4)$, $n \in [2, 12]$.

Remark 3.3. The table representation of the denumerant function cannot be created automatically within *Wolfram Alpha*. It is the use of a circular reference that at each computational step preserves previously tabulated values. Although a complicated *Maple* program can probably create such a table, the discussion of this level of programming is beyond the scope of this book.

3.4 Using 3-D Spreadsheet Modeling in Problem Posing

Problem posing has been on the mathematics education agenda for a long time appearing in different didactic forms [Getzels and Jackson, 1962; Krutetskii, 1976; Watson and Mason, 2005; Cai *et al.*, 2015]. The goal of educational problem posing was to enrich one's learning experience through investigating mathematical ideas, exploring conjectures, and solving problems. Already in the early 1990s, the importance of problem posing for the development of students' mathematical thinking and the growth of educational applications of computers at the pre-college level have been linked together in the standards for teachers: "technology may be used to enhance and extend mathematics learning and teaching ... in the areas of problem posing and problem solving ... [allowing] students to design their own explorations and create their own mathematics" [National Council of Teachers of Mathematics, 1991, p. 134]. The first most commonly known examples of using technology for posing mathematical problems included the development of conjectures in dynamic geometry environments [e.g., Yerushalmy *et al.*, 1993]. More recently, in Japan, "students from different parts of the country [were encouraged to take advantage of technology] to pose problems to each other or communicate their solutions using e-mail, bulletin boards, or video conferencing" [Takahashi *et al.*, 2006, p. 259]. The use of a spreadsheet enables many areas of pre-college mathematics to be explored from a problem-posing perspective. However, the ease of generating data using a spreadsheet brings about new challenges to technology-enabled mathematics pedagogy.

3.4.1 *A notion of problem coherence*

In the context of mathematics teacher education, the use of technology in problem posing can be characterized as a cultural support of teachers' ability to design new curriculum materials for a mathematics classroom. It is cultural because one is encouraged to use the tools of technology developed by advanced members of a technological culture for various practical and scholastic purposes. In particular, this is the case with a spreadsheet, the use of which can be seen as support in the sense that, in the specific context of problem posing, teachers learn how to put to work its computational power in order to formulate a grade-appropriate problem. In some cases, a spreadsheet generates solutions to a problem that is about to be posed and the task for a teacher is to recognize this relation between the two sides of the same coin, problem posing and problem solving. This implies that problem posing and problem solving can be inherently linked to each other through the use of a spreadsheet.

In order for a spreadsheet to have a positive effect on problem posing, one should not only know how to use it but, more importantly, how to interpret numeric data that the software tool generates. Just as "in the workplace ... sophisticated analyses are available at the click of a mouse ... [thus requiring one] to be more mathematically competent in order to understand and interpret the information produced by these analyses" [Advisory Committee on Mathematics Education, 2011, p. 4], learning "the craft of task design" [Conference Board of Mathematical Sciences, 2012, p. 65] in the mathematics teacher education classroom of the digital era requires appreciation of what may be called the didactical coherence of a problem [Abramovich and Cho, 2008]. This notion comprises three distinct components: numerical coherence, contextual coherence and pedagogical coherence. *Numerical coherence* of a problem refers to its formal solvability within a given number system. *Contextual coherence* of a problem refers to a context within which problem posing occurs. Besides the need to understand the context of a problem statement, it requires the appreciation of hidden assumptions grounded into one's real-life experience and cultural background. *Pedagogical coherence* of a problem refers to such issues as attention to students' on-task behavior, the absence of (or minimizing) extraneous data, semantic/contextual clarity, grade appropriateness, the number of possible solutions, and a method of solution expected. Attending to the notion of didactical coherence in the context of problem posing with technology, teacher candidates learn how to help their own students "to identify important quantities in a practical situation ... interpret their mathematical results in the context of the situation and reflect on whether the results make sense, possibly improving the model if it has not served its purpose" [Common Core State Standards, 2010, p. 7].

3.4.2 *Modeling ambiguous problems using a spreadsheet*
Consider the following well-known problem [e.g., New York State Education Department, 1998, p. 77] that can motivate the need for using a spreadsheet in problem reformulation to satisfy the conditions of problem coherence.

Problem 3.3. *A census taker asked the farmer the ages of his three daughters. The farmer told him that the product of their ages is 72 and the sum of their ages is the house number. The census taker performed some computations and then looked at the house number. At this point, he told the farmer that he still could not tell the ages of the daughters. The farmer said, "I forgot to tell you that the oldest likes chocolate pudding." This helped the census taker and he now knew the ages of the three daughters. What are the ages of the three daughters?*

The mathematical structure of Problem 3.3 is based on a possibility of the existence of two (or more) sets of three integers with the same sum and the same product. The task is, given the product, to overcome the hurdle of the multiplicity of answers by using context of the problem. While Problem 3.3 could be solved without using a spreadsheet (indeed, $72 = 3 \cdot 3 \cdot 8 = 2 \cdot 6 \cdot 6$ with only the former factorization having the largest factor), an important role that technology can play in the teaching of mathematics is a possibility of its use to assess students' comprehension of a method. To create a similar problem with two (or more) triples of factors of a whole number having the same sum is not easy. Trial and error approach to this task would require a complete list of three factors of a given number. And this is a three-dimensional problem: to solve the equation $xyz = n$ under certain conditions for the variables x, y, and z; namely, to have (at least) two solutions (x_1, y_1, z_1) and (x_2, y_2, z_2) such that $x_1 + y_1 + z_1 = x_2 + y_2 + z_2$, where $x_1 \leq y_1 < z_1$ and $x_2 < y_2 = z_2$. Thus the main reason to use a spreadsheet in the context of Problem 3.3 is to equip teacher candidates with a tool enabling them to generate multiple problems of the same type that can then be used to assess their students' comprehension of a method and corresponding problem-solving skills.

Using techniques discussed in the context of Problems 3.1 and 3.2, two distinct ways of generating solutions to Problem 3.3 utilizing a spreadsheet as a three-dimensional modeling tool will be presented below. The first way (technologically less challenging) is to add the third dimension to a spreadsheet by assigning one of the variables, say, z, to a slider-controlled cell, and display the sums of three factors of a given number step by step. Another way is to display all possible sums of the three factors of a given number on a single template. This requires the use of a circular reference in a spreadsheet formula.

The spreadsheet shown in Figs. 3.5 and 3.6 is programmed as follows. In the ranges [C3:I3] and [B4:B8] natural numbers (that stand for x and y, respectively) are defined. The slider-controlled cells B3 and A1 are assigned, respectively, for the variable z and the value of a product of three ages. In cell C4 the spreadsheet formula

$\qquad$=IF(AND(C$3>=$B4,$B$3>=C$3,

$\qquad$C$3*$B4*$B$3=$A$1),C$3+$B4+$B$3," ")

is defined and replicated to cell I8. The combination of the logical functions IF and AND makes it possible to consider factors (ages) without regard to order (that is, ensuring that $z \geq x \geq y$). Through such modeling, the above-mentioned triples (3, 3, 8) and (2, 6, 6) having the same sum ($2+6+6=3+3+8=14$) can be found through altering the content of cell B3. The spreadsheet of Fig. 3.5 shows the latter case having the content of cell B3 equal to six. The spreadsheet of Fig. 3.6 shows the former case having the content of cell B3 equal to eight. This concludes the description of the first approach to spreadsheet modeling of Problem 3.3.

The spreadsheet can also retain the sums of ages as the value of cell B3 varies (from the highest to lowest) beginning from the number 72 — a hypothetically the largest age of one of the three daughters (contextually incoherent problem situation). To this end, consider the equation $x + y + z = n$. Given the value $z = z_0$, the pair (x_0, y_0) may not satisfy this equation with a different value of the variable z; that is, the relations $x_0 + y_0 + z_0 = n$, and $x_0 + y_0 + z_1 = n$ cannot hold true simultaneously. This brings about the idea of presenting numbers in a table (range [C4:J8], Fig. 3.7) as the variable z changes in cell B3. In order to retain the sums corresponding to different values of variable z (provided that variables x and y have been generated in the ranges [C3:J3] and [B4:B8] respectively), one has to use a circular reference. To this end, one has to define in cell C4 the spreadsheet formula

=IF(AND(C$3>=$B4, $B$3>=C$3,C$3*$B4*$B$3=$A$1),

C$3+$B4+$B$3, IF($B$3<=$A$1,C4," "))

that includes reference to cell C4 (circular reference) and then replicate it to cell J8. As a result, the spreadsheet displays the number 14 two times on a single template (Fig. 3.7) thus improving the visual component of a computational approach to Problem 3.3.

◢	A	B	C	D	E	F	G	H	I
1	72								
2									
3		6	1	2	3	4	5	6	7
4		1							
5		2						14	
6		3			13				
7		4							
8		5							

Fig. 3.5. Finding the sums step by step with old results disappearing.

◢	A	B	C	D	E	F	G	H	I
1	72								
2									
3		8	1	2	3	4	5	6	7
4		1							
5		2							
6		3		14					
7		4							
8		5							

Fig. 3.6. Another triple of factors of 72 with the same sum.

	A	B	C	D	E	F	G	H	I	J
1	72									
2										
3		72	1	2	3	4	5	6	7	8
4		1	74	39	28	23		19		18
5		2		22	17	15		14		
6		3			14	13				
7		4								
8		5								

Fig. 3.7. Displaying all the sums of three possible factors of 72 simultaneously.

3.4.3 *Using spreadsheets to pose problems*

As was mentioned above, one avenue of problem posing by teachers deals with reformulation of existing problems in order to provide students with a wealthy problem-solving practice. This section shows how a spreadsheet can be used as a generator of similar problems satisfying the conditions of problem coherence. Often, as one takes into account the above-mentioned issues of problem coherence, a change of numeric data, being computationally trivial in the context of a spreadsheet, is not trivial didactically. To illustrate, consider another case of the census taker problem. The triples (1, 9, 32) and (2, 4, 36) do have the same product, 288, and the same sum, 42, thus manifesting numerical coherence of the corresponding problem about census taker and farmer's daughters. Yet, contextually, as the ages of the farmer's daughters, these triples are not realistic (though biologically not impossible).

Similarly, given the product 144 of the ages of three daughters and the sum of ages as the (unknown) house number, there are two pairs of triples, {(3, 6, 8), (4, 4, 9)} and {(3, 4, 12), (2, 8, 9)}, being appropriate as the ages of siblings (Fig. 3.8). The first pair of triples each has the sum 17 as $3 + 6 + 8 = 4 + 4 + 9 = 17$. The second pair of triples has the sum 19 as $3 + 4 + 12 = 2 + 8 + 9 = 19$. While all four triples have the same product, $3 \cdot 6 \cdot 8 = 4 \cdot 4 \cdot 9 = 3 \cdot 4 \cdot 12 = 2 \cdot 8 \cdot 9 = 144$, only the triple (4, 4, 9) satisfies the problem situation in the context of having twins. Therefore, one has to carefully use a spreadsheet to generate both numerically and contextually coherent problems of that kind. After all these considerations, the following problem can be formulated.

Problem 3.4. *A census taker asked the farmer the ages of his three daughters. The farmer told him that the product of their ages is 144 and the sum of their ages is the house number. The census taker performed some computations and then looked at the house number. At this point, he told the farmer that he still could not tell the ages of the daughters. The farmer said, "I forgot to tell you that two of my daughters are twins". This helped the census taker and he now knows the ages of the three daughters. What are the ages of the three daughters?*

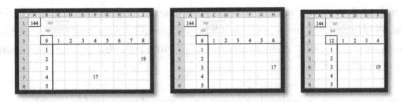

Fig. 3.8. Generating two pairs of triples with same product (144) and sum (17 and 19).

3.4.4 *Formulating TITE problems through parameterization*

The spreadsheet of Fig. 3.3 allows one to formulate a variety of problems leading to a system of three equations in three unknowns. For example, the triple of numbers (1, 2, 1) located in the cells E3, C6, and E6, respectively, can be used to pose

Problem 3.5. *Jonny spent $12 at a book sale where books were priced $2, $3, and $4. He bought twice as many $3 books as $2 books and one fewer $4 books than $3 books. How many books of each type did he buy?*

A mathematical model for Problem 3.5 is the following system of simultaneous equations

$$2x + 3y + 4z = 12, \quad y = 2x, \quad z = y - 1. \tag{3.3}$$

Note that problem solvers are expected to find the single solution of Eqs. (3.3) — $x = 1$, $y = 2$, $z = 1$ — by purely algebraic means.

However, Problem 3.5 is not technology-immune as one can enter into the input box of *Wolfram Alpha* the system of equations (3.3) to get its solution. A more advanced version of this tool, *Wolfram Alpha/Pro*, can present all symbolic computations leading to this solution. In that way, in the digital era, Pólya's adage "It is easy to teach students the right answers; the challenge is to teach students to ask the right questions" [cited in Conference Board of the Mathematical Sciences, 2012, p. 56] acquires new meaning and brings about new challenges for mathematics educators. With this in mind, one can modify Problem 3.5 by not including the exact amount of money spent by Jonny and making this amount an unknown parameter defined by an inequality. For example, one can formulate

Problem 3.6. *Jonny was at a book sale where books were priced $2, $3, and $4. He bought twice as many $3 books as $2 books and one fewer $4 books than $3 books. How many books of each type did he buy if he spent not more than $50?*

This problem is a technology-immune/technology-enabled problem for one cannot use *Wolfram Alpha* to obtain the final answer relying only on "an *automatic transport phenomenon* ... [when the outcome of problem solving process depends on whether] one can feed all the problem's data into the machine" [Guin and Trouche, 1999, p. 205, italics in the original]. Yet the program can be used to support necessary computations. To this end, one can introduce a parameter n into the problem to have the system of equations

$$2x + 3y + 4z = n, \ y = 2x, \ z = y - 1. \qquad (3.4)$$

When system (3.4) is entered in the input box of *Wolfram Alpha*, the program yields (among other things) the solution shown in Fig. 3.9.

Integer solution:

$$n = 4(4m + 3), \quad x = m + 1, \quad y = 2(m + 1), \quad z = 2m + 1, \quad m \in \mathbb{Z}$$

Fig. 3.9. Solving system (3.4) using *Wolfram Alpha*.

Now, using the results provided by *Wolfram Alpha*, one has to solve the inequality $n \le 50$ or $4(4m + 3) \le 50$ whence $m \le 2.375$. That is, $m = 0, 1, 2$. The corresponding values of n, x, y, and z are given by the following quadruples, respectively: $(12, 1, 2, 1), (28, 2, 4, 3), (44, 3, 6, 5)$. This means that under the conditions of Problem 3.6, Jonny can spend not more than $50 as follows: a) $12 by buying one $2 book, two $3 books, and one $4 book; b) $28 by buying two $2 books, four $3 books, and three $4 books; c) $44 by buying three $2 books, six $3 books, and five $4 books.

3.5 Extension into 4-D Modeling

Consider the following variation of Problem 3.5.

Problem 3.7. *How many ways can Jonny spend $25 to buy books priced at $10, $9, $8, and $7?*

In order to create a spreadsheet for modeling this problem, a new computational environment can be created allowing one to deal with partitioning a number into four summands. To begin, consider the Diophantine equation

$$a_1 x + a_2 y + a_3 z + a_4 k = n \qquad (3.5)$$

which describes a partition of n into positive integer summands a_1, a_2, a_3, a_4. The first step is to note that in Eq. (3.5) the inequality

$$k \le INT(n / a_4) \qquad (3.6)$$

holds true. Indeed, let us assume that $k > n / a_4$. Under this assumption, the right-hand side of Eq. (3.5) can be estimated as follows

$$a_1 x + a_2 y + a_3 z + a_4 k > a_1 x + a_2 y + a_3 z + a_4 (n / a_4) = a_1 x + a_2 y + a_3 z + n > n$$

and, therefore, Eq. (3.5) is not satisfied for $k > n / a_4$; this contradiction implies $k \le n / a_4$. Because k is an integer, the last inequality implies even a stronger estimation, namely, $k \le INT(n / a_4)$.

Now, inequality (3.6) allows one to reduce Eq. (3.5) to the family of equations

$$a_1 x + a_2 y + a_3 z = n - a_4 k \qquad (3.7)$$

where $k \in \{0, 1, 2, ..., INT(n / a_4)\}$.

For each such value of k, Eq. (3.7) represents a three-variable equation that can be modeled using a three-dimensional spreadsheet like the one shown in Fig. 3.1. In finding upper boundaries for variables x and

y in Eq. (3.5), one should take into account that these boundaries depend on the value of the right-hand side of (3.7), that is, on the value of $n_k = n - a_4k$, where $k = 0, 1, 2, ..., INT(n/a_4)$. In other words, for each such value of k the inequalities $x \leq INT(n_k/a_1)$ and $y \leq INT(n_k/a_2)$ determine upper boundaries for the variables x and y.

For example, the four-variable equation
$$10x + 9y + 8z + 7k = 25, \qquad (3.8)$$
due to the inequality $k \leq INT(25/7) = 3$, can be reduced to the family of the following four three-variable equations $10x + 9y + 8z = 25$ ($k = 0$), $10x + 9y + 8z = 18$ ($k = 1$), $10x + 9y + 8z = 11$ ($k = 2$), and $10x + 9y + 8z = 4$ ($k = 3$) each of which can be modeled within a three-dimensional spreadsheet shown in Fig. 3.1. Furthermore, a modification of the three-dimensional spreadsheet allows for the full automation of the process of calculating the total number of solutions for any four-variable Diophantine equation of the form (3.2). The programming of such a spreadsheet is discussed elsewhere [Abramovich and Cho, 2008]. However, nowadays, using a spreadsheet for modeling Problem 3.7 is not necessary because of the availability of *Wolfram Alpha* which, in response to the quest "solve over integers $10x + 9y + 8z + 7k = 25$, $x \geq 0$, $y \geq 0$, $z \geq 0$, $k \geq 0$" yields the following three whole number solutions to Eq. (3.6), namely: (0, 1, 2, 0), (0, 2, 0, 1), (1, 0, 1, 1). Note that although permissible values for k are 0, 1, 2, and 3, only the first two values provide solutions to Eq. (3.6). Nonetheless, Jonny can buy any of the four books because within the above three quadruples no zero appears three time at the same position. One can be asked to find the largest sum of money smaller (or the smallest sum larger) than \$25 for which at least one of the books may not be purchased. Such exploration can also be characterized as a TITE problem for it cannot be immediately solved by *Wolfram Alpha* at the push of a button even after carefully entering all data into its input box.

Chapter 4

Prime Numbers

4.1 Introduction

In Chapter 1, different integer sequences were generated using a spreadsheet. It was demonstrated that such sequences can be modeled within a spreadsheet provided that a rule through which their terms develop can be formally described either in the language of mathematics or in the language of computing. For example, some sequences (like the natural number sequence) were defined recursively; for other sequences (like the strings of numbers forming cycles) only different computational procedures were developed.

In this chapter, using a spreadsheet, a new type of integers with the following distinct property will be generated: they cannot be expressed as a product of other integers. Numbers that fall into this category are called prime numbers. For example, 2, 3, 5, and 7 cannot be expressed as a product of smaller integers, whereas 4, 6, 8, and 9 can. In other words, a prime number has exactly two different divisors: one and itself. In mathematics and, in particular, in number theory (known as its Queen [Beiler, 1964]), prime numbers have long been occupying an important place at the cornerstone of many theoretical developments associated with work of great mathematical minds. In the words of Gauss[d] [1966], "The problem of distinguishing prime numbers from composite numbers and of resolving the latter into their prime factors is known to be one of the most important and useful in arithmetic" (p. 396). Some basic ideas about prime numbers have found their place in school mathematics also. As mentioned in the Japanese mathematics curriculum for grades 7-9, "students are expected to understand that natural numbers greater than 1 can be sorted into those numbers whose factors are 1 and the numbers themselves, and those with more factors — that is, prime numbers and non-prime numbers" [Takahashi *et al.*, 2006, p. 213]. The difficulty in identifying prime numbers and discovering prime factors of composite

---

[d] Carl Friedrich Gauss (1777-1855, Germany) — one of the greatest mathematicians in the history of mankind.

numbers increases as the magnitude of numbers increases. This is an obstacle for extending the students' learning about prime numbers beyond the basic ideas. Nonetheless, the study of prime numbers in school mathematics is relevant because "the prime factorization of natural numbers [being unique] is equivalent to factoring polynomials" [*ibid*, p. 213], an equivalence through which many topics in high school algebra can be conceptualized in terms of generalized arithmetic.

The initial understanding of the prime vs. composite notion, expected from students in Japan (and elsewhere in the world) does not help one to decide what kind of number, say, 131 is. One may wonder: Can be this (or any other) number plugged into a formula to get an answer? (Of course, just as in many other cases, a powerful computational engine *Wolfram Alpha* can give the answer immediately. Yet, it does not explain the answer to support one's conceptual understanding). Unfortunately, unlike in the case of natural, odd and even numbers, as well as more complicated numeric sequences such as polygonal numbers (Chapter 9), no formula is known to decide whether a number is a prime or to produce a prime number given its rank. A call for such a formula, whatever its meaning [Matiyasevich, 1999], was probably motivated by the fact, famously proved by Euclid[e] using a proof by contradiction, that there are infinitely many prime numbers. In his proof, Euclid made an assumption about the existence of the largest prime number p and then constructed an example of another prime number greater than p, thus running into a contradiction with his own assumption.

Another historically famous fact associated with prime numbers is due to Euler[f] who constructed the function $f(n) = n^2 - n + 41$ (alternatively, $f(n) = n^2 + n + 41$ — a form suggested by Legendre[g] [Mollin, 1997]) of an integer variable n that for all $n \in [1, 40]$ (alternatively, for all $n \in [0, 39]$) produces only prime numbers (although not consecutive ones). Furthermore, as n increases, $f(n)$ continues

---

[e] Euclid — the most prominent Greek mathematician of the third century B.C.

[f] Leonhard Euler — a Swiss mathematician of the eighteenth century, the father of all modern mathematics.

[g] Andrien-Marie Legendre — a French mathematician of the eighteenth-nineteenth centuries.

producing mostly prime numbers. For example, in the range $n \in [41, 120]$ more than 75% of the values of $f(n)$ are prime numbers.

In the digital age, despite the absence of a prime number formula, both the model through which prime numbers emerge and the computational algorithm that produces primes within a given range can be provided. In particular, a spreadsheet can be put to work to generate prime numbers and test integers for primality. Note that because *Wolfram Alpha* can verify not only whether a number is a prime or not, but can also generate prime numbers, the purpose of using a spreadsheet in this context is mostly educational and it is aimed at supporting one's conceptual understanding through the discussion of spreadsheet modeling techniques. However, in Sec. 4.6 below a spreadsheet will be used as a tool capable of fast-testing different prime-producing polynomials of the second degree and higher and identifying the lengths of the strings of prime numbers that such polynomials produce. In that case, the use of a spreadsheet extends the boundaries of purely educational computations and takes over the capability of *Wolfram Alpha*.

4.2 Generating Prime Numbers Within a Spreadsheet

A simple procedure of generating consecutive prime numbers was devised more than two thousand years ago by Eratosthenes[h]. Commonly referred to as the Sieve of Eratosthenes, this procedure allows one to obtain all the prime numbers less than any given integer N, by crossing out from the set of all natural numbers greater than one and less than N the multiples of each of the primes up to $\sqrt{N}$ in turn. All numbers that remain undeleted are the primes sought. Suppose that all multiples of the primes 2, 3, 5, ..., p_n, where p_n is the n-th prime number, were eliminated. Then, only the multiples of primes greater than p_n and not divisible by any of the first n primes have survived the elimination process. The smallest multiple of p_{n+1} that was not eliminated by the first n primes is p_{n+1}^2. Indeed, the multiples of p_{n+1} smaller than p_{n+1}^2 are $2p_{n+1}, 3p_{n+1}, ..., p_n \cdot p_{n+1}$ and each of them was eliminated by one of the

---

[h] Eratosthenes — a Greek scholar of the third century B.C.

first *n* primes. For example, the smallest composite number that cannot be eliminated by the first three primes — 2, 3, and 5 — is 49, the square of the fourth prime number.

The so described sieve is shown in Fig. 4.1 where numbers in the range [2, 49] that are not crossed out are those that survived the process of elimination by the first four primes — 2, 3, 5, and 7. Indeed, none of these numbers is a *k*-multiple (*k* > 1) of a natural number greater than one. In other words, these numbers satisfy the definition of prime numbers for they cannot be represented as a product of two smaller positive integers.

2	3	4	5	6	7	8	9	10	11	12	13	14	15	16	17
18	19	20	21	22	23	24	25	26	27	28	29	30	31	32	33
34	35	36	37	38	39	40	41	42	43	44	45	46	47	48	49

Fig. 4.1. Eliminating integers with more than two different factors.

Toward the end of using a spreadsheet as a generator of prime numbers, note that the process of eliminating integers with more than two different factors shown in Fig. 4.1 can be represented in the form of a division table (Fig. 4.2) where vertical and horizontal inputs are, respectively, tested numbers *n* and numbers *d* (e.g., 2, 3, 5, 7) that are not multiples of the tested numbers. Therefore, the examination of the quotient *n* / *d* is a two-dimensional computational process, which can be performed in this table for different values of *n* and *d*.

In a course of this process, several cases are possible. First, such a quotient might be either an integer greater than one — in this case *n* is a composite number (for it has more than two different factors, namely, one, *d*, and the quotient) and the process terminates; or it is a non-integer and the process continues. The process terminates, however, when it yields the number one — the case of *n* being a prime number; or an integer greater than one — the case of *n* being a composite number. Note that the seven numbers not crossed out in the first row of the "paper-and-pencil" sieve of Fig. 4.1 can be used to test for primality all whole numbers not greater than 360. Indeed, the first number in such a table (with the first seven primes only) that would be erroneously identified as a prime is 361 (the square of 19) as it is not the multiple of any of the original seven prime numbers used in the division test.

	2	3	5	7	11	13
2	1	**PRIME**				
3	3/2	1	**PRIME**			
4	2	**COMPOSITE**				
5	5/2	5/3	1	**PRIME**		
6	3	**COMPOSITE**				
7	7/2	7/3	7/5	1	**PRIME**	
8	4	**COMPOSITE**				
9	9/2	3	**COMPOSITE**			
10	5	**COMPOSITE**				
11	11/2	11/3	11/5	11/7	1	**PRIME**

Fig. 4.2. The Sieve of Eratosthenes in the form of a two-dimensional table.

Once the Sieve of Eratosthenes is represented in the form of a two-dimensional table, its resemblance to a spreadsheet becomes apparent. The computational capacity of a spreadsheet for two-dimensional modeling makes it possible to use the software as the generator of prime numbers. The next section provides the details of required programming and suggests several follow-up computational activities using a spreadsheet.

4.3 Construction of the Spreadsheet Sieve

The programming of a spreadsheet that generates prime numbers through the Sieve of Eratosthenes, is based on a single formula =IF(MOD(A2,B$1)>0,A2, IF(A2=B$1,A2,0)) (or its equivalent =IF(MOD(A2,B$1)=0, IF(A2=B$1,A2,0), A2)) defined in cell B2 and replicated across columns and down rows (Fig. 4.3). This single formula requires knowledge of at least a few consecutive prime numbers starting from 2 to begin a test for primality. As Fig. 4.2 suggests, if a tested number is not divisible by a prime (taken from the top row of the table where the primes are stored), the former becomes engaged with the next prime in this row; otherwise, depending on whether the quotient equals one or not, the tested number, respectively, is either prime or composite.

Alternatively, one can define in cell B2 the spreadsheet formula =IF(INT(A2/B\$1)*SIGN(A2-B\$1)=A2/B\$1,0,A2), where the function SIGN(x) assumes one of the values 1, 0, or -1 depending on whether $x > 0$, $x = 0$ or $x < 0$, respectively. If a tested number is composite and thus divisible by one of the primes, then the condition of equality in this formula is true and a zero generated in cell B2 terminates the sieve process. Otherwise, if a tested number is either prime or is a survivor of division, the number moves one cell to the right within the spreadsheet-based sieve.

Figure 4.3 shows that tested numbers are stored in column A and prime numbers used in the test are stored in row 1. More specifically, in cell A2 number 2 is defined, in cell A3 the spreadsheet function =1+A2 is defined and replicated down (thus filling column A with consecutive natural numbers), then in row 1 (range B1:L1) the first 11 primes (from 2 to 31) are entered. Thus, setting 31 as the largest prime to be used in the test enables one to generate all prime numbers that do not exceed 1368 ($1368 = 37^2 - 1$, where 37 is the smallest prime number greater than 31).

	A	B	C	D	E	F	G	H	I	J	K	L	M	N	O
1		2	3	5	7	11	13	17	19	23	29	31			95
2	2	2	2	2	2	2	2	2	2	2	2	2	2	2	
3	3	3	3	3	3	3	3	3	3	3	3	3	3	3	
4	4													5	
5	5	5	5	5	5	5	5	5	5	5	5	5	5	7	
6	6													11	
7	7	7	7	7	7	7	7	7	7	7	7	7	7	13	
8	8													17	
9	9	9												19	
10	10													23	
11	11	11	11	11	11	11	11	11	11	11	11	11	11	29	
12	12													31	
13	13	13	13	13	13	13	13	13	13	13	13	13	13	37	
14	14													41	
15	15	15												43	
16	16													47	
17	17	17	17	17	17	17	17	17	17	17	17	17	17	53	
18	18													59	
19	19	19	19	19	19	19	19	19	19	19	19	19	19	61	
20	20													67	

Fig. 4.3. The spreadsheet-based Sieve of Eratosthenes.

In order to improve visualization, a spreadsheet can be set not to display zero values, because in this computational environment zeros play a technical role only. To this end, in the context of Macintosh, one enters the *Excel* menu, choose the line *Preferences* to open its *View* dialog box, set the box *Show Zero Values* blank, and then click OK. In the context of Windows, one enters the *File* menu, chooses *Options* (under Help) and then *Advanced* in the left pane, unchecks the *Show a Zero in Cells That Have Zero Value* in the *Display Options for this Worksheet* section, and clicks OK. In addition, the spreadsheet formula =IF(L2=0, " ", L2) defined in cell M2 and replicated down column M makes it possible to fill column M with consecutive primes that belong to the range of numbers tested. One can test numbers for primality in the strings of 500 numbers starting with 2 and attach a slider with an increment 500 to cell A2. Of course, the choice of the length of such a string is arbitrary.

In that way, the spreadsheet can count the number of primes within a chosen string of natural numbers and then display primes in column N without gaps. To this end, one can define the spreadsheet formula =COUNT(M2:M501) in cell O1 that counts in column M the number of primes within the range [A2:A501] filled with consecutive natural numbers. Now, the syntactic meaning of the formula defined in cell M2 becomes clear. In the case when a cell contains zero, regardless whether it is displayed or not, a computer counts zeros along with the primes. But due to the formula =IF(L2=0, " ", L2) a cell with zero is replaced by an empty cell which is disregarded by the function COUNT. In order to display primes in column N without gaps, one can define in cell N2 the spreadsheet formula

=IF(A2-A\$2+1<=O\$1, SMALL(M2:M501, A2-A\$2+1), " ")

and replicate it down column N (Fig. 4.3). Note that the spreadsheet variable A2-A\$2+1 always yields the positional rank of its first reference in column A. Indeed, in cell N20 this variable becomes A20-A\$2+1 and, in the case of the number 2 in cell A2, it assumes the value of 19 which (being equal to $20 - 2 + 1$) represents the positional rank of the number in cell A20 within the string of consecutive natural numbers with the smallest one located in cell A2.

Changing the content of cell A2 makes the spreadsheet to count and to display in cell O1 the quantity of prime numbers within the corresponding string of consecutive natural numbers. For example, the spreadsheet shows that there are 95 primes among the first 500 natural numbers and 73 primes among the remaining three-digit numbers. The same count of prime numbers can be obtained through *Wolfram Alpha* (see Fig. 4.4). So, the latter tool can be used for verifying the correctness of spreadsheet modeling. However, using the spreadsheet, one can visualize not only the distribution of prime numbers but also search electronically for specific gaps between consecutive primes, something that *Wolfram Alpha* cannot easily provide. By exploring these gaps one can come to understand that the lack of any apparent pattern in the distribution of prime numbers among natural numbers (presented in terms of the gaps) is a reason for the absence of a formula producing infinitely many consecutive primes. Other spreadsheet-enabled explorations involving prime numbers are suggested in the sections that follow.

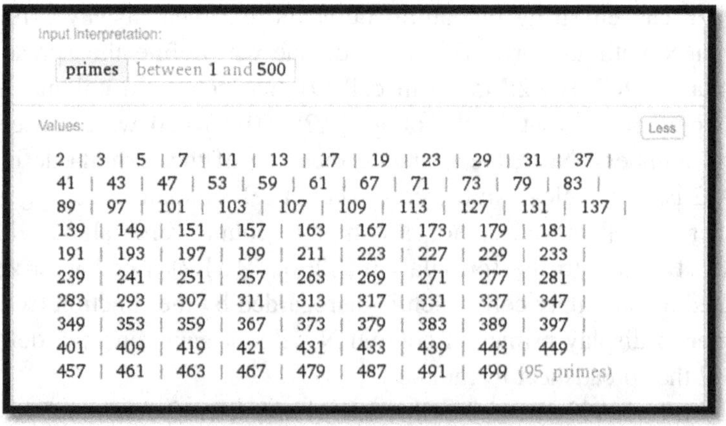

Fig. 4.4. All 95 prime numbers smaller than 500 generated by *Wolfram Alpha*.

Remark 4.1. The spreadsheet of Fig. 4.3 shows the smallest prime factor of a tested number. It does not provide the prime factorization though. It is possible to construct a spreadsheet that generates all prime factors of a given number at the push of a button. However, a much more simple factoring technique is to test for primality the number divided by its smallest prime factor. For example, in the case of the number 1547, the smallest prime factor identified by the spreadsheet is 7; then for the number $\cdot 1547/7$ the smallest factor is 13; finally, for the number $1547/(7 \cdot 13)$ is 17. Thus, the prime factorization is $1547 = 7 \cdot 13 \cdot 17$.

Remark 4.2. Another interesting (though not surprising) fact about the function $f(n) = n^2 - n + 41$ (Sec. 4.1) is that the differences between the successive forty primes — 41, 43, 47, 53, 61, ..., 1523, 1601 — form an arithmetic sequence 2, 4, 6, 8, ..., 78. This is because the first difference of a quadratic function is a linear function which, in turn, is a formal representation of an arithmetic sequence. In particular, $\Delta f(n) = f(n+1) - f(n) = 2n$. One can use a spreadsheet to explore the function $f(n)$. A classic prime number theory result about integer arithmetic sequences with the first term a and difference d is due to Dirichlet[i] who proved that if $d \geq 2$ and $a \neq 0$ are relatively prime (unlike the above-mentioned sequence 2, 4, 6, 8, ...), then there are infinitely many prime numbers among the terms of the sequence $x_n = a + dn$, $n = 0, 1, 2, \ldots$. For example, the sequence 3, 8, 13, 18, 23, 28, 33, 38, 43, ... (sequence A030431 in the On-line Encyclopedia of Integer Sequences® [http://oeis.org]; see also [Van de Walle *et al.*, 2013, p. 15] — quite a different source providing another example of hidden mathematics curriculum of teacher education [Abramovich, 2009]) consists of infinitely many primes, including 3, 13, 23, and 43. In other words, the relation $x_n = a + dn$ may be interpreted as a "formula for prime numbers" [Matiyasevich, 1999, p. 13].

4.4 Twin Primes

A pair of prime numbers with difference two is called twin primes. This concept is famous for being associated with an open problem (by the time of writing this book) whether there exist infinitely many such pairs. Yet, some simple explorations with twin primes are possible in the context of spreadsheets. To this end, note that twin primes fall into three groups depending on their endings (i.e., the last digit): 1 and 3 (11 & 13 is the smallest such pair), 7 and 9 (17 & 19 is the smallest such pair), 9 and 1 (29 & 31 is the smallest such pair). All other last digits — 0, 2, 4, 5, 6, 8 — belong to composite numbers (the digit 5 is the only example of a prime number with the ending 5). Let us consider the group of twin primes with the endings 1 and 3. The first pair in this group is (11, 13). One can use a spreadsheet to generate a few next pairs with these

---

[i] Peter Gustav Lejeune Dirichlet — a German mathematician of the nineteenth century.

endings: (41, 43), (71, 73), (101, 103), and then guess the next pair in this group. One can recognize a pattern among the above four pairs: these primes are in arithmetic progression with the difference 30. This may suggest that the next pair is (131, 133). However, one can see that whereas 131 is a prime number, the number 133 is not. Moreover, the spreadsheet shows (Fig. 4.5) that the number 133 has indeed more than two factors for it does not survive divisibility by seven in the sieve process. That is, the number 133 is a multiple of seven.

	A	B	C	D	E	F	G	H	I	J	K	L	M	N
1		2	3	5	7	11	13	17	19	23	29	31		
2	131	131	131	131	131	131	131	131	131	131	131	131	131	131
3	132													137
4	133	133	133	133										139
5	134													149
6	135	135												151
7	136													157
8	137	137	137	137	137	137	137	137	137	137	137	137	137	163
9	138													167
10	139	139	139	139	139	139	139	139	139	139	139	139	139	173

Fig. 4.5. The number 133 is a multiple of seven.

What is the next pair of twin primes with endings 1 and 3? Does the pair (161, 163) comprise twin primes? One can also find that the number 161 is a multiple of seven. This kind of experience with empirical induction can make one's guessing more careful. Finally, one can find that the 191 and 193 are twin primes. So, the following first five pairs of twin primes with endings 1 and 3 have been found:

(11; 13), (41; 43), (71; 73), (101; 103), and (191, 193).

At that point the following questions may arise:

What is the next pair of twin primes in this group?

Do there exist another four pairs of twin primes in this group with 30 as a common difference?

Is 30 the smallest gap between the pairs of this group?

Do there exist more than four consecutive pairs of twin primes in this group with 30 as a common difference?

Does there exist another gap which occurs three times running in consecutive pairs of twin primes in this group?

What do all the gaps occurring between consecutive pairs of twin primes in this group have in common?

What is the largest gap between two consecutive pairs of twin primes of this group within the range, say, [1:10000]?

Are there formulas through which twin primes in this group can be generated (including composite numbers with difference two)?

Is it possible to find consecutive pairs of twin primes in this group with gaps in arithmetic progression?

The same explorations can be carried out for twin primes with the other two pairs of endings, namely, (7, 9) and (9, 1). For example, one can find several consecutive pairs of twin primes with the endings 7 band 9 starting from the pair (17, 19) and then guess the next pair in this sequence. Likewise, one can find several consecutive pairs of twin primes with the endings 9 and 1, starting from the pair (29, 31) and then guess the next pair in this sequence. Explorations with the endings (7, 9) and (9, 1) differ from the case (1, 3) in the complexity of empirical induction.

4.5 Exploring Gaps Between Consecutive Primes

It is possible to build another computational environment allowing for the study of gaps between primes. For example, one can find that, whereas there are 24 twin primes in the range [2, 500], there exist 25 pairs of consecutive primes with difference four (such pairs are called cousin primes): (7, 11), (13, 17), (19, 23), (37, 41), ..., (487, 491). In search for a pattern among cousin primes, note that only the first three pairs are consecutive prime numbers. This search can also reveal the existence of four pairs of cousin primes being consecutive prime numbers. Extending the search to the range [501, 1000], one can discover four pairs of cousin primes consisting of consecutive prime numbers: (853, 857), (859, 863), (877, 881), and (883, 887).

Also, there exist five sequential primes in arithmetic progression with the difference six: 5, 11, 17, 23, 29. Do there exist another five primes in arithmetic progression with the same property? The new environment would make it possible to search not only for primes in arithmetic progression but also for primes with gaps in arithmetic progression. For example, the primes 347, 349, 353, 359, 367 have gaps in arithmetic progression (2, 4, 6, 8). Are there another five primes with gaps in arithmetic progression? By simply changing the entry of a single cell one can discover many fascinating properties of prime numbers.

Remark 4.3. Note that there is no string of primes in arithmetic progression with gaps also in (a non-stationary) arithmetic progression (i.e., with a non-zero difference). Indeed, such a string of primes has the form p, $p + d$, $p + 2d$, $p + 3d$, … with the gaps between the terms of the string forming a stationary sequence d, d, d, … . Before attempting any search for numbers with specified properties using a spreadsheet, one has to make sure that what is searchable is not obviously contradictory in its characterization. For example, searching for an odd number among even numbers is meaningless as the two sets of numbers $\{2n\}$ and $\{2n - 1\}$ have no common elements by definition. Yet, the search for squares among triangular numbers is a classic problem of number theory solved by Euler who found a closed formula for triangular squares (see Sec. 9.18 of Chapter 9).

4.6 Testing Prime-Producing Polynomials Using a Spreadsheet

As was mentioned in Sec. 4.3, the set of primes in the range [2, 31] can generate an extended set of 218 primes that do not exceed the number 1368 (=$37^2 - 1$, where 37 is the smallest prime number greater than 31). The smallest prime number that is greater than 1368 is 1373. Then, using this set of 218 primes, one can create an environment capable of testing for primality all numbers that do not exceed 1885128 (= $1373^2 - 1$, where 1373 is the smallest prime number greater than 1368). Using the results of this test, the first 218 primes can be augmented to accommodate even greater numbers to be tested for primaliy. Indeed, the square of 1885128 is a number with 13 digits. However, in the actual practice of mathematics education one rarely needs to decide on primality of such a large number. Otherwise, one can use *Wolfram Alpha.*

Figure 4.6 shows that such an environment does not require much space of a spreadsheet to be displayed. Indeed, the extended set can be stored and hidden in column A beginning from cell A3 with a tested number located in cell B2; the formula =IF(MOD(B2,A3)>0,B2,IF(B2=A3,B2,0)) can be defined in cell B3 and replicated down column B; cell C2 contains the formula =IF(B...=0, "NO", "YES"), where B... is a cell immediately below the largest prime used in this test. For example, 5153 is the largest prime number in the set of the first 686 primes; in that case, the last formula would refer to cell B689. Also, this formula generates the YES message in cell C2 each time when a tested number survives the sieve process and, thereby, is a prime number. Note that because the smallest prime number greater than 5153 (the 686th prime) is the number 5167, by using the first 686 primes one can correctly test for primality all numbers smaller than $5167^2 = 26697889$.

◢	A	B	C
1			
2	IS THIS A PRIME?	73939133	YES
3			
4	2	73939133	
5	3	73939133	
6	5	73939133	
7	7	73939133	
688	5147	73939133	
689	5153	73939133	

Fig. 4.6. A spreadsheet-based test for primality.

Another interesting use of a spreadsheet is associated with a classic prime number theory problem. In 1857, Buniakovski[j] conjectured that if a polynomial $f(x)$ with integer coefficients is irreducible over the integers (i.e., cannot be factored into two non-constant polynomials) and $GCD(f(1), f(2), f(3),...) = 1$, where GCD is the greatest common divisor, then $f(x)$ produces an infinite number of primes (http://mathworld.wolfram.com/BouniakowskyConjecture.html). While such a polynomial has not been found (yet), there are many prime-producing polynomials in addition to the Euler polynomial mentioned above in Sec. 4.1) of the second degree and higher that generate strings of prime numbers of different length. The spreadsheet shown in Fig. 4.6

---

[j] Viktor Buniakovski (sometimes spelled Bouniakowsky) — a Russian mathematician of the nineteenth century.

can be extended to enable exploration of different prime-producing quadratics and other polynomials listed in [Pegg, 2006].

	A	B	C	D	E	F
1					40	1
2	IS THIS A PRIME?	1681	NO	LENGTH FOUND IN E1	41	
3				1681		
4	2	1681	1	1		1
5	3	1681	2	1		-1
6	5	1681	3	1		41
7	7	1681	4	1		
688	5147		685			
689	5153		686			

Fig. 4.7. Testing the Euler's polynomial $f(x) = x^2 - x + 41$.

	A	B	C	D	E
1					57
2	IS THIS A PRIME?	5141923	NO	LENGTH FOUND IN E1	57
3				5141923	
4	2	5141923	0	1	
5	3	5141923	1	1	
6	5	5141923	2	1	
7	7	5141923	3	1	
688	5147	0	684		
689	5153	0	685		

Fig. 4.8. Polynomial (4.1) generates a 57-long string of prime numbers.

Figure 4.7 shows such an extension where different quadratics of the form $ax^2 + bx + d$ can be tested for primality. To this end, the values of the coefficients a, b, and d can be entered in cells F4, F5, F6 (named a, b, d), respectively; the formula a*E2^2+b*E2+d — in cell B2 (duplicated in cell D3) in order to generate values of the corresponding quadratics, the values of x are controlled by a slider attached to cell E2; in column C beginning from cell C4 (controlled by a two-value slider {0, 1}) consecutive integers starting from either 0 or 1 are defined; in cell D4 the formula =IF(F$1=0,0,IF(C4=E$2,1,D4)) including a circular reference is defined and replicated down column D to allow for counting prime numbers generated by the corresponding quadratics; in cell C2 the formula =IF(B689=0,"NO","YES") which tests a number from cell B2 for primality is defined; in cell E1 the formula

=IF(C2="NO",SUM(D4:D100)-1," ") generates the length of the corresponding string of prime numbers immediately after cell B2 generates the first composite number; and in cell D2 the formula =IF(C2="NO","LENGTH FOUND IN E1"," ") is defined to indicate that the length sought has been found. Using such a spreadsheet, one can see that when $a = 1$, $b = -1$, $d = 41$ (the Euler polynomial), 40 primes emerge one by one as x varies from 1 to 40. Likewise, other quadratics can be tested for the length of the strings of prime numbers that they produce.

A notable polynomial, producing a string of 57 prime numbers for all $x \in [0, 56]$, has the form

$$f(x) = 0.25(x^5 - 133x^4 + 6729x^3 - 158379x^2 \atop +1720294x - 6823316). \tag{4.1}$$

Polynomial (4.1), according to Pegg [2006], appears to be a producer of the longest string of prime numbers (known by the time of writing this book). The spreadsheet of Fig. 4.8 confirms the length 57 (cell E1) after cell B2 is entered by the formula (duplicated in cell D3) defining polynomial (4.1) with x being replaced by cell reference E2. Because x is varying from zero, the spreadsheet is set to display zero values.

Remark 4.4. An interesting task that can be carried out in the context of a spreadsheet integrated with *Maple* is to prove by the *Maple*-based method of mathematical induction (see Sec. 1.7 of Chapter 1) that polynomial (4.1) assumes an integer value for any natural number x. Put another way, one has to prove that the sum

$$P(n) = n^5 - 133n^4 + 6729n^3 - 158379n^2 + 1720294n - 6823316$$

is divisible by four for any $n \in N$.

As shown in Fig. 4.9, just as in the second illustration of Sec. 1.7, the proof requires two steps (after making an obvious conclusion that $P(1) = -5254804$ is a multiple of four; indeed, the last two digits of $P(1)$ make a trivial multiple of four). The first step is to develop the difference $P(n+1) - P(n)$ which is not obviously divisible by four for any $n \in N$ and therefore, the inductive assumption that $P(n)$ is divisible by four does not imply the same for $P(n+1)$. In other words, passing from n to $n + 1$ does not demonstrate what is to be proved. So, the second step it to prove by the same method of mathematical induction that the difference $P(n+1) - P(n)$ is divisible by four. Indeed, this difference turns out to be equal to $20n^3 - 1536n^2 + 37252n - 278216$, the polynomial with all the coefficients being multiples of four. That is, assuming that $P1(n) = 5n^4 - 522n^3 + 19399n^2 - 297098n + 1568512$ is divisible by four and then obtaining the identity $P1(n + 1) - P1(n) = 20n^3 - 1536n^2 + 37252n - 278216$ imply that $P1(n + 1)$ is divisible by four as well, thereby completing the inductive transfer from n to $n + 1$.

$P(n) := n^5 - 133\,n^4 + 6729\,n^3 - 158379\,n^2 + 1720294\,n - 6823316$
$$n \to n^5 - 133\,n^4 + 6729\,n^3 - 158379\,n^2 + 1720294\,n - 6823316$$
$P(n+1) - P(n)$
$(n+1)^5 - 133\,(n+1)^4 + 6729\,(n+1)^3 - 158379\,(n+1)^2 + 1720294 - n^5 + 133\,n^4 - 6729\,n^3$
$+ 158379\,n^2$
$simplify(\%)$
$$5\,n^4 - 522\,n^3 + 19399\,n^2 - 297098\,n + 1568512$$
$P1(n) := 5\,n^4 - 522\,n^3 + 19399\,n^2 - 297098\,n + 1568512$
$$n \to 5\,n^4 - 522\,n^3 + 19399\,n^2 - 297098\,n + 1568512$$

$>\ P1(n+1) - P1(n)$
$\quad 5\,(n+1)^4 - 522\,(n+1)^3 + 19399\,(n+1)^2 - 297098 - 5\,n^4 + 522\,n^3 - 19399\,n^2$
$>\ simplify(\%)$
$$20\,n^3 - 1536\,n^2 + 37252\,n - 278216$$

Fig. 4.9. *Maple*-based mathematical induction proof regarding polynomial (4.1).

Remark 4.5. Complicated spreadsheet programming can be used to generate other numbers with special divisibility properties such as perfect, deficient, and abundant numbers. However, by using integrated spreadsheets this programming can be replaced by *Wolfram Alpha* in generating those numbers. At the same time, some basic features of integers such as having exactly two different divisors can be explored in a spreadsheet environment for educational purposes. As the teaching of simple algorithms of arithmetic continues having been taught nowadays in a traditional way, such operations as extracting square and cube roots (something that was taught in the past) are not taught in the schools any more and, instead, outsourced to calculators and computers. For example, using the spreadsheet of Fig. 4.5, one can immediately see why the number 133 is a composite one and the number 131 is not. The full explanation of the latter requires understanding that if 131 has survived divisibility by the primes not greater than 11, there is no need to include larger primes in the test because the smallest composite number divisible by a prime greater than 11 is 169 (the square of 13).

4.7 Fermat Primes and Euler's Factorization Method

A classic example of the necessity of rigor in mathematics deals with Fermat[k] primes, numbers with long and interesting history. Gauss linked these numbers to the ancient problem of constructing a regular polygon with compass and straightedge and proved at the age of 19 that if n is a Fermat prime, then a regular n-gon can be so constructed [Conway and Guy, 1996]. As was mentioned above, no formula exists to produce a prime number of any given rank. In search for such a formula, Fermat came across the following five numbers represented through the powers of two:

$$F_0 = 2^{2^0} + 1 = 3, F_1 = 2^{2^1} + 1 = 5, F_2 = 2^{2^2} + 1 = 17,$$

$$F_3 = 2^{2^3} + 1 = 257, F_4 = 2^{2^4} + 1 = 65537.$$

All the numbers 3, 5, 17, 257, and 65537 turned out to be primes. From here, Fermat conjectured that all numbers of the form $F_n = 2^{2^n} + 1, n = 0, 1, 2, \ldots$, are prime numbers.

Using the spreadsheet of Fig. 4.6 (alternatively, *Wolfram Alpha*), one can check that all the five numbers are indeed primes. From here one can guess (like Fermat did) that the pattern continues with the growth of the powers of two in the expression for F_n. However, this is not true: the next Fermat number $F_5 = 2^{2^5} + 1$ is composite. In the digital era, it is pretty easy to come to this conclusion. Indeed, entering the formula $= 2^{32} + 1$ in cell B2 (Fig. 4.6) yields the message NO in cell C2 meaning that the number did not survive divisibility by 641. Next, entering $= (2^{32} + 1) / 641$ (which equals to 6700417) into cell B2 results in the message YES, thus confirming the primality of the resulting quotient. Therefore, $F_5 = 641 \cdot 6700417$ is a composite number and this result refutes the conjecture by Fermat. Note that the exploration with Fermat primes can be carried out in the context of *Wolfram Alpha* without any

---

[k] Pierre de Fermat — a French mathematician of the seventeenth century.

difficulty — in response to the request "factor 2^32+1" the program immediately generates the two non-trivial divisors of F_5. However, the above prime factorization was known long before the digital era. This is a truly remarkable gem of number theory and one may want to know how it was developed. The prime factorization of the number F_5 is due to Euler who found the two factors theoretically. Introducing a historical component into the modern context of technology use enables one to see the development of mathematical ideas made possible by connecting different concepts of mathematics.

4.7.1 *A numeric example*

Euler's factorization method is applicable to integers that can be represented as the sums of two squares in more than one way. To explain this method, consider first a numeric example of factoring the number 85. We have (see Sec. 6.2 of Chapter 6) $85 = 9^2 + 2^2 = 7^2 + 6^2$. From here it follows that $9^2 - 7^2 = 6^2 - 2^2$ or, after factoring, $(9-7)(9+7) = (6-2)(6+2)$. Now, consider the differences $9 - 7 = 2 = 2 \cdot 1$ and $6 - 2 = 4 = 2 \cdot 2$ for which $GCD(2,4) = 2$ and, consequently, $GCD(1, 2) = 1$. The equality $2 \cdot 1 \cdot (9+7) = 2 \cdot 2 \cdot (6+2)$ can be simplified to the form $9 + 7 = 2 \cdot (6 + 2)$ whence $9 + 7 = 16 = 4 \cdot 4$ and $6 + 2 = 8 = 4 \cdot 2$. Now, one can factor 85 as follows:

$$85 = \frac{1}{4}(2 \cdot 9^2 + 2 \cdot 2^2 + 2 \cdot 7^2 + 2 \cdot 6^2)$$

$$= \frac{1}{4}[(9+7)^2 + (9-7)^2 + (6+2)^2 + (6-2)^2]$$

$$= \frac{1}{4}[16^2 + 2^2 + 8^2 + 4^2] = \frac{1}{4}[(2 \cdot 8)^2 + (2 \cdot 1)^2 + (8 \cdot 1)^2 + (2 \cdot 2)^2]$$

$$= \frac{1}{4}(2^2 \cdot 8^2 + 2^2 \cdot 1^2 + 8^2 \cdot 1^2 + 2^2 \cdot 2^2) = \frac{1}{4}[8^2 \cdot (2^2 + 1^2) + 2^2 \cdot (2^2 + 1^2)]$$

$$= \frac{1}{4}[(2^2 + 1^2)(8^2 + 2^2)] = \frac{1}{4} \cdot 5 \cdot 68 = 5 \cdot 17.$$

4.7.2 *Generalization*

In general, let us consider an odd (the only even prime number is 2) integer N such that $N = a^2 + b^2 = u^2 + v^2$. Then $a^2 - u^2 = v^2 - b^2$ from where it follows, after factoring differences of the squares, that $(a-u)(a+u) = (v-b)(v+b)$. Let $GCD(a-u, v-b) = k$, then $a-u = k \cdot l$ and $v-b = k \cdot m$ where $GCD(l, m) = 1$.

Now, the equality $(a-u)(a+u) = (v-b)(v+b)$ can be replaced by the equality $k \cdot l \cdot (a+u) = k \cdot m \cdot (v+b)$ whence $l \cdot (a+u) = m \cdot (v+b)$. The last relation implies $a+u = n \cdot m$ and $v+b = n \cdot l$. This allows one to factor N as follows:

$$N = \frac{1}{4}(2a^2 + 2b^2 + 2u^2 + 2v^2)$$

$$= \frac{1}{4}[(a+u)^2 + (a-u)^2 + (b+v)^2 + (b-v)^2]$$

$$= \frac{1}{4}(n^2m^2 + k^2l^2 + n^2l^2 + k^2m^2)$$

$$= \frac{1}{4}[n^2(m^2 + l^2) + k^2(m^2 + l^2)]$$

$$= \frac{1}{4}(k^2 + n^2)(m^2 + l^2).$$

4.7.3 *Spreadsheet implementation of Euler's factorization method*
The first step in applying the above factorization method to the number $F_5 = 2^{32} + 1$ is to find the first representation of F_5 as a sum of two squares. To this end, in the spreadsheet of Fig. 4.10 cell A1 is entered with the number F_5, the range [A2:A1001] is filled with the first thousand natural numbers a, the range [B2:B1001] with their squares a^2, the range C2:C1001 with the (positive) differences $F_5 - a^2$. In column D the search for an integer b such that $F_5 - a^2 = b^2$ is carried out. The spreadsheet of Fig. 4.10 carries out the first step. One can see that the first (trivial) representation of F_5 as the sum of two squares is $(2^{16})^2 + 1^2$. The following formulas defined in the corresponding cells can be used to program the spreadsheet.

(A1)$\rightarrow$ = 2^32+1. Cell A2 is slider-controlled and its is set to display numbers congruent to one modulo 1000 by entering the number 1000 as the page change into the slider's Format Control box (with 30000 as the maximum value). Next, (A3)$\rightarrow$ =A2+1 — replicated to cell

A1001; (B2)→ = A2^2 – replicated to cell B1001; (C2)→ =IF(A$1-B2>0,A$1-B2," ") — replicated to cell C1001.

(D2)→ =IF(OR(C2=" ",C2<B2)," ", IF(SQRT(C2)=INT(SQRT(C2)), SQRT(C2)," ")) — replicated to cell D1001; (E2)→ =IF(COUNT(D2)>0,1," ") — the formula counts the number of non-empty cells in column D; (F2)→ =COUNTIF(D2:D30000,">0") — this formula counts the number of cells with non-zero values in column D.

The next two formulas defined in cells G2 and H2 include the function LOOKUP (Sec. 1.6.1 of Chapter 1): (G2)→ =IF(SUM(E2:E1001)=0," ",LOOKUP(1,E2:E1001,D2:D1001)) — this formula searches for the number 1 in the range [E2:E1001] and returns the value from the range [D2:D1001] sharing the row with 1 (that is, it returns the integer value of b satisfying the equation $F_5 - a^2 = b^2$); (H2) → =IF(SUM(E2:E1001)=0," ",LOOKUP(1,E2:E1001,A2:A1001)) — this formula is looking for the number 1 in the range E2:E1001 and returns the value from the range A2:A1001 that is in the same row with 1 (that is, it returns the value of a satisfying the equation $F_5 - a^2 = b^2$). Finally, we have (I2)→ =IF(G2=" "," ",IF(G2^2+H2^2=A1,"YES"," ")) — this formula verifies the first representation of F_5 as a sum of two squares and confirms it with the affirmative message YES. Note that such a verification would not be possible without the function LOOKUP capable of picking up the values of a and b based on information associated with these values, yet presented outside the ranges allocated for them.

The second step is to find another representation of F_5 as a sum of two squares. To this end, the already found values of a and b have to be entered in cells A1005 and A1007 (Fig. 4.11) through, respectively, the formulas

(A1005)→ =IF(COUNT(D2)>0,G2,A1005) and

(A1007)→ =IF(COUNT(D2)>0,SQRT(A1-A1005^2),A1007).

The two formulas include circular references to allow for the values of a and b to stay without change as one changes the value of cell A2 through the slider attached to it. The goal of using the slider is to find another representation, $F_5 = u^2 + v^2$, among the values of u greater than 1000. The need for a and b to be available concurrently with u and v is due to the need to compute the values of $a \pm u$ and $v \pm b$ once u and v have been located. Eventually (see Sec. 7.4.2), one needs to compute $k^2 + n^2$ and $l^2 + n^2$ where $k = GCD(a-u, v-b)$, $l = (a-u)/k$, $m = (v-b)/k$, and $n = (v+b)/l$. This explains the rest of the formulas used in programming the spreadsheet of Fig. 4.11.

$(A1009) \rightarrow =A1;$
$(B1005) \rightarrow =IF(D2=G2," ",IF(COUNT(F2)>0,G2,C1005));$
$(B1007) \rightarrow =IF(B1005=" "," ",SQRT(A1-B1005^2));$
$(B1009) \rightarrow =IF(B1005=" "," ", (F1004/2)^2+(F1007/2)^2);$
$(C1005) \rightarrow =IF(B1005=" "," ",A1005-B1005);$
$(C1007) \rightarrow =IF(B1005=" "," ",B1007-A1007);$
$(C1009) \rightarrow =IF(B1005=" "," ", F1006^2+F1005^2);$
$(D1005) \rightarrow =IF(B1005=" "," ",A1005+B1005);$
$(D1007) \rightarrow =IF(B1005=" "," ", A1007+B1007);$
$(F1004) \rightarrow =IF(B1005=" "," ",GCD(C1005,C1007));$
$(F1005) \rightarrow =IF(B1005=" "," ", C1005/F1004);$
$(F1006) \rightarrow =IF(B1005=" "," ", C1007/F1004);$
$F(1007) \rightarrow =IF(B1005=" "," ",D1007/F1005).$

	A	B	C	D	E	F	G	H	I
1	4294967297								
2	1	1	4294967296	65536	1	1	65536	1	YES
3	2	4	4294967293						
4	3	9	4294967288						

Fig. 4.10. Locating the first sum of two squares, $a^2 + b^2$.

	A	B	C	D	E	F
1004	*a*	*u*	*a-u*	*a+u*	*k*	8
1005	65536	62264	3272	127800	*l*	409
1006	*b*	*v*	*v-b*	*v+b*	*m*	2556
1007	1	20449	20448	20450	*n*	50
1008	*N*	*x*	*y*			
1009	4294967297	641	6700417			

Fig. 4.11. Completing prime factorization of $2^{32} + 1$ after locating u and v.

Chapter 5

Mathematical Concepts as Modeling Tools

5.1 Introduction

The problem formulated in this section motivates the development of several number theory concepts to be used as tools in computing applications. This problem occurred in the course of constructing a spreadsheet-based manipulative-computational environment for exploring percentage problems in the mathematics classroom. Such explorations are in support of various standards for teaching school mathematics around the world that, due to "a widespread inability [within workforce] to understand percentages" [Cockcroft Report, 1982, p. 8], aim to ensure that students know how to "interpret percentages and percentage changes as fraction or a decimal ... [and] compare 2 quantities using percentages" [Department for Education, 2013b, p. 39]. Because these topics "are notoriously difficult for students in early adolescence" [Expert Panel on Student Success in Ontario, 2004], the standards, emphasize the importance of "expressing one quantity as a percentage of another" [Ministry of Education, Singapore, 2006, p. 6], using knowledge of "ratios and proportionality to solve a wide variety of percent problems" [Common Core State Standards, 2010, p, 46], and "understanding of proportional relationships using percent, ratio, and rate" [Ontario Ministry of Education, 2005a, p. 99]. Teacher candidates, already at the elementary level, should know "that finding a percentage part of a quantity is equivalent to finding a fractional part of the quantity" [Ministry of Education, Singapore, 2012, p. 24] and understand how to "calculate percentages mentally and write equations to show the algebra behind the mental methods" [Conference Board of the Mathematical Sciences, 2012, p. 29]. Towards this end, the pedagogic intent of the author, while preparing to address didactical expectations of the above type, was to utilize a spreadsheet as a virtual manipulative with an interactive link to numeric notation to solve percentage problems. In such an environment, a student is presented with a rectangular grid which comprises fewer than 100 cells. On this grid a number of cells is shaded (Fig. 5.1) so that the shaded part constitutes a whole number percentage of the entire grid. This condition is always true on a 100-cell grid; on a smaller grid it is not necessarily satisfied. For example, 15 cells represent 30% of a 50-cell grid; yet 15 represent 37.5% of a 40-cell grid. Let 15 cells on a 50-cell grid be shaded as shown in Fig. 5.1. A task for a student is to evaluate what percent of the grid is shaded and enter this percentage number into the answer box (a cell of the spreadsheet). An interactive link established between iconic and numeric notations of the environment enables a spreadsheet to display a message evaluating the content of the answer box. If the student's answer is incorrect, this message suggests continuing the task on an identical adjacent non-shaded grid originally hidden from view. The objective of this new task is to give a student an opportunity to appropriate an incorrect answer as a thinking device and, in doing so, to shade a region on the adjacent grid that does correspond to this (incorrect) answer. In other words, the didactic emphasis of the task is to prevent undesirable consequences of a negative evaluation and to allow for the latter to awaken new meanings for a student in the context of a non-authoritative pedagogy.

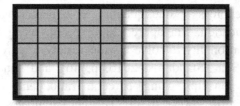

Fig. 5.1. What percent of the grid is shaded?

5.2 Arithmetical Properties of Rectangular Grids

Technically, the idea of turning a negative evaluation into a generator of new meanings cannot be implemented in this environment without constrains. Indeed, not every whole number percentage can be represented by a shaded part of a grid with fewer than 100 cells. For example, 30% of a 50-cell grid constitute 15 cells (Fig. 5.1) whereas 25% of the same grid constitute 12.5 cells. Yet, in a spreadsheet environment, only an entire cell may be shaded. So, if a student evaluates the shaded part of a 50-cell grid as 25%, the task of shading 25% of the identical grid would be impossible. In such a way, a dichotomy between the non-authoritative pedagogy of a task and its semiotic structure brings about the following problematic situation: The environment must have means to narrow down a possible guessing by a student.

One way of resolving this situation is to create an environment offering a few choices of percentage numbers, including the correct choice, from which to select an answer. For example, four such choices on a 40-cell grid may be: 10% (4 cells), 25% (10 cells), 55% (22 cells), and 75% (30 cells). In turn, new questions can be formulated. Among them:

Are there more such choices available on a 40-cell grid?

What are these choices?

How many ways can one shade a 40-cell grid in a whole number percentage?

What could the choices be on a 30-cell grid and how many are there?

Do there exist rectangular grids for which multiple choices cannot be found?

What is special about such grids?

Given an n-cell grid, $n \leq 100$, what is the total number of choices available on this grid?

Initial attempts to find answers to these questions bring about

Problem 5.1. *On an n-cell grid, $0 < n \leq 100$, k cells are shaded. What percent of the grid is shaded? For which n and k is this percent a whole number?*

This context-bounded problem can serve as a springboard into the development of concepts with an instant link to situational referents provided by a manipulative component of the environment. For the sake of correct formulation of the problems that follow, the inequality $n \leq 100$ limiting the size of a grid will be assumed for all the tasks dealing with rectangular grids. Note that limiting our considerations to rectangular (rather than concave polygonal) grids is immaterial from the mathematical perspective.

Although Problem 5.1 is formulated in algebraic terms, the sample of concrete, numeric problems used for introducing the problematic situation of the manipulative environment can mediate one's grasp of this general situation for which familiar numeric examples serve as situational referents. To this end, two steps in solving the problem can be suggested. The first step is to construct the ratio k/n and multiply it by 100%. This yields the answer to the first question of Problem 5.1 in the following form: *(k/n)100% of the grid is shaded.* But could it be physically done? This depends on whether n divides $100k$ or not. Therefore, the second step is to introduce the following constraint: the shaded region must constitute a whole number percentage of the entire grid. Note that some integral pairs (k, n), $0 < k \le n$, make the ratio $(k/n)100$ a whole number whereas other such pairs do not. For example,

$k = 2$ and $n = 3$ yield $\frac{2}{3} \cdot 100 = 66.(6)$ — a repeating decimal. Yet $k = 3$

and $n = 5$ yield $\frac{3}{5} \cdot 100 = 60$ — a whole number. This empirical approach

to the second part of Problem 5.1 prompts using a spreadsheet to generate pairs (k, n) satisfying either the equality $INT(\frac{100k}{n}) = \frac{100k}{n}$ or $MOD(100k, n) = 0$. In turn, the results of such use of a spreadsheet can be incorporated into the design of the manipulative-computational environment for exploring percentage problems in the context of a non-authoritative pedagogy.

5.3 Exploiting Syntactic Versatility of a Spreadsheet

A spreadsheet's computational capacity dealing with processes depending on two variables (demonstrated in Chapters 2 and 4) makes it possible to evaluate the expression $\frac{100k}{n}$ numerically by computing its values for every pair of whole numbers k and n within a specified range. In search of whole number percentages such computing involves the division of two integers, and a whole number quotient resulted from this division can be interpreted twofold: (i) division with a zero remainder, and (ii) division with a zero fractional part. As far as a spreadsheet-based tool kit of computing devices is concerned, these two ways of generating whole number percentages can be implemented by using, respectively, either the function MOD or the function INT. Whereas the former computes a remainder from the division of two numbers which is supposed to be zero, the latter truncates the fractional part of the quotient which, in turn, has to be the fixed point of the greatest integer function. A pedagogy utilizing syntactic versatility of a spreadsheet facilitates the process of mastering a formal mathematical symbolism through the recognition of the availability of different syntactic means. This pedagogy contributes to a classroom culture in which learners of mathematics are granted flexibility and freedom in developing various problem-solving strategies.

When one deals with the notation $MOD(5, 2)$ in a pencil-and-paper environment, the resulting number is up to his or her decision; when one enters the formula =MOD(5, 2) into a cell of a spreadsheet, the resulting number is due to the spreadsheet. Once the spreadsheet generates the result $MOD(5, 2) = 1$, it enables one's reflection on an abstract notation through *numerical evidence* — a persuasive argument which is both heuristic and demonstrative [Beth, 1966]. The same interplay exists between the notation $INT(5 / 2)$ and the spreadsheet formula =INT(5/2). Using either of the spreadsheet functions MOD or INT from its tool kit of computing devices makes it possible to display only integer values of the expression $\dfrac{100k}{n}$.

Figure 5.2 shows a fragment of the spreadsheet with percentage numbers generated within the 100×100 range of the variables k and n. Row 2 beginning from cell C2 and column B beginning from cell B3 are filled with positive integer values n and k, respectively; cell C3 is entered with the spreadsheet formula

=IF(AND(C$2>=$B3,INT($A$2*$B3/C$2)=$A$2*$B3/C$2),
$A$2*$B3/C$2, " "),

or its alternative

=IF(AND(C$2>=$B3,MOD($A$2*$B3,C$2)=0), $A$2*$B3/C$2, " ").

Either formula is replicated across the spreadsheet. Both formulas contain an absolute reference to cell A2 the content of which can be instantly changed resulting in an interactive change of numeric data across the spreadsheet.

Numerical evidence of whole number percentages provided by spreadsheet modeling calls for interpretation of visual information. This interpretation may include: (i) creating a problematic situation by asking a meaningful question; (ii) conjecturing and debating an answer to this question, and (iii) providing a formal demonstration (proof) of the conjecture. With this in mind, the following questions aimed at in-depth investigation of arithmetic properties of rectangular grids can be formulated.

What is special about a 20-cell grid on which 20 different choices to shade in a whole number percentage are possible?

What is special about a 9-cell grid on which only one (trivial) choice is possible?

How can one distinguish between the numbers 20 and 9 in that sense?

Has it something to do with the numbers being of different parity?

And if so, why does a 5-cell grid allow for five choices?

What does the number of non-empty cells in each column depend on?

Attempts to answer these questions give rise to a new problem formulated here both in context-bounded and algebraic forms.

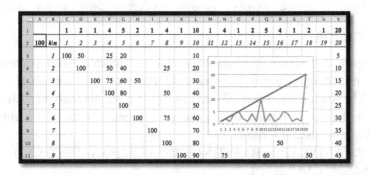

	A	B	C	D	E	F	G	H	I	J	K	L	M	N	O	P	Q	R	S	T	U	V
1			1	2	1	4	5	2	1	4	1	10	1	4	1	2	5	4	1	2	1	20
2	100	k\n	1	2	3	4	5	6	7	8	9	10	11	12	13	14	15	16	17	18	19	20
3		1	100	50		25	20					10										5
4		2		100		50	40			25		20										10
5		3			100	75	60	50				30										15
6		4				100	80			50		40										20
7		5					100					50										25
8		6						100		75		60										30
9		7							100			70										35
10		8								100		80					50					40
11		9									100	90		75			60			50		45

Fig. 5.2. Numeric and graphic solutions to Problem 5.2.

Problem 5.2. *How many subgrids of an n-cell grid constitute its whole number percentage part, provided that shape and location of a subgrid is not important? Put another way, given a positive integer n, how many ways can one choose a whole number k, $k \le n$, such that $100 \cdot k / n$ is a whole number also?*

Computer-generated data shown in Fig. 5.2 makes it possible to approach the problem from a visual perspective. Indeed, if $n = 5$ all whole numbers k from 1 through 5 work for Problem 5.2, whereas the case $n = 7$ allows for one solution only, namely $k = 7$. What is special about the number 7 that allows for only one (trivial) solution? What if $n = 32$, 88 or 96? That is, how many subgrids of, respectively, the 32, 88, and 96-cell grids can be located to give a whole number percentage part of a whole grid?

Further utilization of the spreadsheet pictured in Fig. 5.2 can be helpful in exploring Problem 5.2 both from numeric and graphic perspectives. To this end, the number of choices for each grid in the range [1,100] can be found electronically: the use of the statistical function COUNT in row 1 makes it possible to display the total number of choices for a specified grid. In addition, when data in rows 2 and 1 (the sizes of the grids and the corresponding number of ways of making $100 \cdot k / n$ a whole number), are considered, respectively, as x and y values, a graphic representation of their relationship can be constructed. A chart embedded into the spreadsheet template (Fig. 5.2) shows such a graph oscillating between the lines $y = 1$ and $y = x$.

5.4 From Data-Driven Conjecturing to Proving

The use of a spreadsheet makes it possible to introduce Problem 5.2 in four different notations: iconic, algebraic, numeric, and graphic. This variety of representational formats highlights the semiotic heterogeneity of a spreadsheet-based tool kit. The tool kit metaphor refers to an array of semiotically heterogeneous representational formats which mediate one's mathematical thinking in a technology-rich environment. The basic tenet of the tool kit approach is that qualitatively different representational formats promote one's conceptual development through mediated action and mathematical visualization, which "is a foundation to the development of abstract understanding, confidence and fluency" [Western and Northern Canadian Protocol, 2008, p. 10].

In particular, the use of a spreadsheet-based tool kit can stimulate a classroom discourse aimed at finding the solution to Problem 5.2. The following questions concerning the behavior of the oscillating graph shown in the chart of Fig. 5.2 can support the discourse.

How can this graph be interpreted?

What are the points where the graph and the line $y = x$ coincide?
What are the points where the graph reaches its minimum?
How many different y-values does the graph have within the range [1, 20]?
What is special about these values?
The following note by a teacher candidate reflected the summary of this discourse and pointed to a solution to Problem 5.2. As she put it:

"The most unexpected discovery [we made in the class] was that the number of cells that result in whole number percentages is equal to the greatest common divisor between the original number and 100."

It is the numerical evidence of the computational environment that visually mediated the inductive discovery mentioned by the teacher in her comment about the course taught by the author. In turn, the availability of a plausible conjecture enabled the course participants to grasp and carry out its proof with a relative ease. The corresponding deductive argument (proof) can be presented as follows.

Let $n = rs$ and $100 = ms$, where $GCD(r,m) = 1$ and $GCD(n,100) = s$.

Therefore, $\dfrac{100k}{n} = \dfrac{msk}{rs} = \dfrac{mk}{r}$. As $k \le n$, we have $MOD(mk, r) = 0$ for

all $k \in \{r, 2r, ..., sr\}$. This implies that there are exactly $s = GCD(n,100)$ different choices to shade an n-cell grid in the whole number percentage. This completes the proof.

Remark 5.1. The approach of using mathematical concepts as tools in computing applications is in agreement with a Japanese course of study mathematics that suggests, "the greatest common divisor and least common multiple should be dealt with in line with concrete situations without putting too much emphasis on the formality of obtaining them" [Takahashi *et al.*, 2004, p. 245]. The approach supports the notion that "mathematical understanding is the ability to justify ... where a mathematical rule comes from" [Common Core State Standards, 2010, p. 4]. The above classroom episode shows that "instruction in mathematics should always foster critical thinking ... investigate and make conjectures about mathematical concepts ... employ *inductive* *reasoning*, ... use *deductive reasoning* to assess the validity of conjectures and to formulate proofs" [Ontario Ministry of Education, 2005b, p. 13, italics in the original].

5.5 Emergence of a New Concept Through De-contextualization of Meditational Means

5.5.1 *Subgrids of the second order*

An emphasis on solving problems in the context of spreadsheet modeling brings about an effective approach to the development of mathematical concepts. The word effective is used here to mean that at any stage of one's concept awareness the approach makes it possible to link a concept to its situational referent so that the concept can then be mediated by action using semiotic heterogeneity of a spreadsheet. Then, through reflection on his or her action, one can make a cognitive leap from the plane of action to the plane of thought where decontextualization of mediational means leading to generalization becomes possible.

Still, the process of generalization is rooted in a particular observation. Consider all subgrids of a 56-cell grid each of which constitutes its whole number percentage part. According to Problem 5.2, there exist $GCD(100, 56) = 4$ such parts that we can call subgrids of the first order. Yet, the 56-cell grid can itself be considered as a whole number percentage part of a larger grid (e.g., an 80-cell grid as $\frac{56}{80} \cdot 100\% = 70\%$) and the subgrids of the former (the 56-cell grid) may, in turn, be considered as parts of the latter (the 80-cell grid) that we can call the subgrids of the second order. For example, a 28-cell grid, being a subgrid of the first order for a 56-cell grid (28 is 50% of 56), is also a subgrid of the second order for an 80-cell grid (28 is 35% of 80) for which the 56-cell grid is 70%. Such a modification imparted to Problem 5.2 can become a starting point for extended inquiries and new discoveries. The following situations can then be explored.

How does the number of such subgrids of a given grid change if they are considered as the subgrids of the second order? For example, a 56-cell grid is 70% of an 80-cell grid (Fig. 5.3).

How many subgrids of the 56-cell grid constitute a whole number percentage part of the 80-cell grid? In other words, how many subgrids of the second order of the 80-cell grid each of which contains not more than 56 cells constitute its whole number percentage part, provided that shape and location of such subgrid is not important? (One such a (28-cell) subgrid of the second order is shaded dark in Fig. 5.3). The last question gives rise to a new problematic situation referred to below as Problem 5.3.

Fig. 5.3. The dark part as a subgrid of the second order of the 80-cell grid.

5.5.2   *"Real" and "hypothetical" subgrids*

Before the discussion of Problem 5.3, consider one more example. Let a 10-cell grid be referred to as 15% of another, an x-cell grid. Does such larger grid exist? Put another way, is x a whole number? We have $\frac{15}{100}x = 10$ whence $x = \frac{1000}{15} = \frac{200}{3}$ — not a whole number. Therefore, a 10-cell grid is 15% of a "hypothetical" grid with a non-whole number, $200/3$, of cells. This numeric example points to the fact that a whole number may constitute a whole number percentage of a fraction (e.g., 10 is 15% of the fraction $200/3$). In what follows, the discussion will be extended to include two cases: one that deals with a grid that constitutes a whole number percentage of a "real" grid; and another that deals with a grid which constitutes a whole number percentage of a "hypothetical" grid. In such a way, one can formulate

Problem 5.3. *How many subgrids of an n-cell grid constitute a whole number percentage part of a larger, perhaps, a "hypothetical" grid for which the n-cell grid is a p% part, provided that shape and location of a subgrid is not important?*

For example, as shown in Fig. 5.3, the 56-cell grid, for which the 28-cell grid is a subgrid, is 70% of the 80-cell grid ($\frac{56}{80} \cdot 100\% = 70\%$);

and the 28-cell grid is 35% of the 80-cell grid ($\frac{28}{80} \cdot 100\% = 35\%$). Put another way, as $80 = \frac{56 \cdot 100}{70}$, we have $\frac{28}{80} \cdot 100 = \frac{28 \cdot 100}{56 \cdot 100 / 70} = \frac{70}{2} = 35$.

In general, given n and p, one has to find $k, k \le n$, from the following conditions: $n = x \cdot \frac{p}{100}$ and $INT(\frac{k}{x} \cdot 100) = \frac{k}{x} \cdot 100$. As $x = n \cdot \frac{100}{p}$, one has to find k such that the expression $\frac{k \cdot 100}{100 \cdot (n/p)} = \frac{kp}{n}$ is a whole number. In other words, given whole numbers p and n, Problem 5.3 seeks the number of ways one can choose a whole number k in order to make the expression $\frac{k}{n} \cdot p$ a whole number.

A slight modification of the spreadsheet of Fig. 5.2 makes it possible to model a solution to Problem 5.3 (by making p a variable) as it turns into Problem 5.2 when $p = 100$. The spreadsheet shown in Fig. 5.4 represents such a modified environment in which cell A2 contains the value $p = 15$. One can then ask:

What are the numbers displayed in row 1 (Fig. 5.4)?

By analogy with Problem 5.2, one can come with the following

Conjecture. *The numbers in row 1 are the greatest common divisors between the number 15 and the numbers in row 2.*

In other words, in row 1 the spreadsheet counts the number of subgrids of a given n-cell grid (n is fixed) each of which constitutes a whole number percentage p (displayed in cell A2) of a larger, perhaps, hypothetical, grid.

To prove this conjecture, in other words, "in order to verify that properties observed in several examples may be true in general" [Takahashi *et al.*, 2006, p. 47], one has to show that there exist

$GCD(n, p)$ values of k such that the expression $\dfrac{k}{n} \cdot p$ is a whole number.
Let $p = ms$ and $n = rs$, where $s = GCD(n, p)$. Then

$$\frac{k}{n} \cdot p = \frac{k \cdot (ms)}{rs} = \frac{k}{r} \cdot m.$$

By analogy with the case of a 100-cell grid, due to the inequality $k \le n = rs$, the values of $k \in \{r, 2r, 3r, ..., sr\}$ make the expression $\dfrac{k}{r} \cdot m$ a whole number. Therefore, on an n-cell grid which constitutes $p\%$ of another grid there exist $GCD(n, p)$ subgrids of the second order each of which constitutes a whole number percentage of the latter grid. In other words, given whole numbers n and p, there exist $GCD(n, p)$ ways to choose a whole number $k \le n$ in order to make the expression $\dfrac{k}{n} \cdot p$ a whole number. This statement, mediated by the spreadsheet of Fig. 5.4, gives a solution to Problem 5.3.

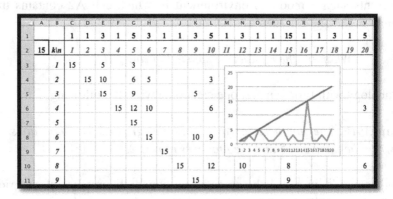

Fig. 5.4. Numeric and graphic solutions to Problem 5.3.

5.6 The Sum of the Greatest Common Divisors as a Problem-Solving Tool

When n and p are fixed, the value of $GCD(n, p)$ gives the total number of the second order subgrids related to two grids — an n-cell grid and an x-cell grid where $x = \dfrac{n}{p} \cdot 100$. One can make n a variable and consider all possible second order subgrids related to all possible grids that constitute $p\%$ of other grids, including hypothetical ones. For example, 56 is 70% of 80, 49 is 70% of 70, and 40 is 70% of $400/7$. We have $GCD(56,70)=14, GCD(49,70)=7$, and $GCD(40,70)=10$. Therefore, the total number of subgrids of the second order of the 56, 49, and 40-cell grids each of which constitutes 70% of other grids is 31 (= 14 + 7 + 10).

With this in mind, one can allow for the alteration of both k and n in the expression $\dfrac{k}{n} \cdot p$ while keeping p constant. Such extension of Problem 5.3 leads to another problem situation decontextualized from grids and their subgrids. Before formulating a new problem, note that not only $k \le n$ but, if one wants to limit the discussion to the grids not exceeding 100 cells (see Sec. 5.1), the inequality $n \le p$ should be assumed as well. Indeed, if $n > p$, then for the larger, x-cell grid of which the n-cell grid is $p\%$ we have $x = \dfrac{100n}{p} > \dfrac{100p}{p} = 100$.

Problem 5.4. *Given a whole number p, how many ways can one choose a positive integer pair* $(k,n), k \le n \le p$, *so that* $\dfrac{k}{n} \cdot p$ *is a whole number?*

In search of a solution to Problem 5.4 one can denote $T(p)$ a numerical function of a positive integer p which counts the number of all pairs $(k,n), k \le n \le p$ such that $\dfrac{k}{n} \cdot p$ is a whole number. In order to compute the function $T(p)$ for particular values of p, one can modify the environment used for modeling Problem 5.3 to allow for the summing of numbers in row 1 (Fig. 5.4). For example, when the entry of cell A2 is 15, $T(15)$ — the sum of the first 15 numbers in row 1 — equals 45. This fact can be interpreted twofold: (i) in iconic form — all 15% parts of all, perhaps hypothetical, grids have the total of 45 (second order) subgrids each of which constitutes a whole number percentage of the corresponding grid; (ii) in algebraic form — there exists 45 whole number pairs $(k, n), k \le n \le p$, such that $\dfrac{15k}{n}$ is a whole number.

In general, the numerical function $T(p)$ can be expressed as follows

$$T(p) = GCD(p, 1) + GCD(p, 2) + \ldots + GCD(p, p).$$

According to Problem 5.3, given whole numbers p and n, there exist $GCD(n, p)$ values of k such that $\dfrac{k}{n} \cdot p$ is a whole number. Therefore, the function $T(p)$ gives the total number of such pairs (k, n), $k \le n \le p$. In other words, $T(p)$ gives a solution to Problem 5.4. In particular, the total number of subgrids of all rectangular grids with not more than 100 cells each of which constitutes its whole number percentage part is 520. This number gives the total number of choices to shade in whole number percentage all such rectangular grids. In such a way, the emergence of a sense-making concept, the sum of greatest common divisors, from a non context-bounded task, like (purely algebraic) Problem 5.4, is an example of how meaning of a mathematical concept may stem from a problem to which it provides a mean of solution [Vergnaud, 1982].

Remark 5.2. The remaining sections of this chapter deal with explorations of the function $T(p)$ drawing on the Euclidean algorithm as a means of finding the GCD of two integers. One may note that in the presence of a spreadsheet function capable of finding the GCD of two or more (in fact, up to 255) whole numbers at the push of a button, the use of the Euclidean algorithm is not necessary. Indeed, using this function, the values of $T(p)$ can be computed within a two-column table augmented by a single cell that adds up those values within a specified p-range (cell C1 and columns A and B in Fig. 5.8). However, similarly to the case of using a spreadsheet-based sieve of Eratosthenes (Chapter 4), the utilization of the Euclidean algorithm for the construction of computational learning environments shown in Figs. 5.5-5.7 has mostly a didactical merit. As will be shown below, besides the historical significance of the algorithm, its development and computational utilization allow for the emergence of, perhaps unexpected, connections to other notable concepts of mathematics.

5.7 Revisiting Familiar Concepts in New Environments

5.7.1 *Improving computing complexity*

The formula defining the function $T(p)$ is not a constructive formula; that is, it is not applicable for the straightforward analytical or graphical investigation of the behavior of the function. A special case of a prime number p can be explored with a relative ease yielding a simple formula for $T(p)$. Indeed, if p is a prime number then the last term in the sum $GCD(p,1) + GCD(p,2) + ... + GCD(p, p)$ equals p whereas each of the remaining $p-1$ terms equals one. Therefore, when p is a prime number we have $T(p) = (p-1)\cdot 1 + p = 2p - 1$. For a composite number p, because for $i < p$ at least one value of $GCD(p, i)$ is greater than 1, the value of $T(p)$ can be estimated from below as follows: $T(p) > 2p - 1$.

A spreadsheet can be used to investigate the behavior of the function $T(p)$ in the case of a composite p. However, such numerical investigation becomes a cumbersome task within the existing computational environment of Fig. 5.4. There are two reasons for that. First, the task requires a manual repetitive alteration of the content of cell A2 (an input p) in order to evaluate the output $T(p)$. Second, the evaluation of a single value of $T(p)$ requires p^2 calculations. In such a way, a new problematic situations dealing with the issue of improving computing complexity of an environment can be formulated. The questions to be answered are:

How can the values of $T(p)$ be generated without the need to manually alter the content of cell A2 and store each value one by one?

How can one reduce the number of calculations in evaluating the value of $T(p)$?

Note that the issue of computing complexity becomes meaningful for the learners of mathematics if, indeed, the environment enables an improvement in terms of reducing the number of calculations and the way the calculations could be performed. In that way, a spreadsheet becomes an amplifier of mathematical activities. Through participation in the improvement of an algorithm one can gain a better understanding of mathematical relationships and concepts that underpin a computational procedure. As the next two sections will show, a dynamic representation of a non-constructively defined function $T(p)$ and the utilization of the Euclidean algorithm of finding the greatest common divisor of two integers are the answers to the problematic situations dealing with the improvement of computing complexity. Furthermore, through the improvement of a computational procedure for exploring the function $T(p)$, an unexpected connection of the Euclidean algorithm to Fibonacci numbers and the appearance of prime numbers as a numeric solution to an equation involving the function $T(p)$ will emerge in the collateral learning format. Once again, in resolving these situations through spreadsheet modeling familiar mathematical concepts can be revisited through their use as problem-solving tools of computational environments. This will give to the concepts new meanings and tie them to a social dimension of knowledge acquired through activities mediated by tools which are the products of social evolution and cultural development.

5.7.2 *Linking the Euclidean algorithm to Fibonacci numbers*

The Euclidean algorithm is one of the topics included in the undergraduate mathematics curriculum for secondary teacher candidates [Conference Board of the Mathematical Sciences, 2012]. This algorithm, used in finding the $GCD(a, b)$ of two whole numbers a and b, is a recursive procedure of successive divisions based on a relation of the fundamental importance in number theory — the remainder (R) equals the dividend (D) minus the divisor (d) times the quotient (Q). That is, $R = D - dQ$.

To clarify, consider a numeric example of dividing 6 into 20. In this process, the following four numbers can be identified: 20 is the dividend, 6 is the divisor, 3 is the quotient, and 2 is the remainder. Moreover, the four numbers satisfy the equality $20 = 3 \cdot 6 + 2$ whence $2 = 20 - 3 \cdot 6$. In other words, the remainder 2 is equal to the dividend 20 minus the product of the quotient 3 and the divisor 6. This definition of remainder is integrated into the $MOD(a, b)$ function that returns the remainder of a divided by b.

Setting two whole numbers a and b as dividend and divisor, one can find their remainder, which (if greater than zero) becomes a new divisor, whereas the old divisor becomes a new dividend. This process of calculating remainders continues until a zero remainder is reached. At that point, the process terminates, as division by zero is not possible. And, according to Euclid, the last non-zero remainder represents the $GCD(a, b)$.

As an example, let us find the $GCD(132, 93)$ using the Euclidean algorithm. To this end, one can consider the following steps required to reach a zero remainder.

Step 1. Divide 93 into 132 and find the remainder R_1: $R_1 = 132 - 1 \cdot 93 = 39$. Put another way, $R_1 = MOD(132, 93) = 39$.

Step 2. Divide R_1 into 93 and find the remainder R_2: $R_2 = 93 - 2 \cdot 39 = 15$. That is, $R_2 = MOD(93, 39) = 15$.

Step 3. Divide R_2 into R_1 and find the remainder R_3: $R_3 = 39 - 2 \cdot 15 = 9$. That is, $R_3 = MOD(39, 15) = 9$.

Step 4. Divide R_3 into R_2 and find the remainder R_4: $R_4 = 15 - 1 \cdot 9 = 6$. That is, $R_4 = MOD(15, 9) = 6$.

Step 5. Divide R_4 into R_3 and find the remainder R_5: $R_5 = 9 - 1 \cdot 6 = 3$. That is, $R_5 = MOD(9, 6) = 3$.

Step 6. Divide R_5 into R_4 and find the remainder R_6: $R_6 = 6 - 2 \cdot 3 = 0$. That is, $R_6 = MOD(6, 3) = 0$.

The process terminates as $R_6 = 0$. The sequence of remainders generated through the above six steps is:

$$R_1 = 39, R_2 = 25, R_3 = 9, R_4 = 6, R_5 = 3, R_6 = 0.$$

According to Euclid, the last non-zero remainder, R_5, in this sequence is equal to $GCD(132, 93)$. Indeed, as $GCD(6, 3) = 3$ (according to the expression for R_6), it follows that 3 is a divisor of 6 and 9 (according to the expression for R_5), and therefore, 3 is a divisor of 9 and 15 (according to the expression for R_4). From here it follows that 3 divides 39 and 15, then 3 divides 93 and 39, and finally, 3 divides 132 and 93. If there exists number $n > 3$ that divides both 132 and 93, this value of n divides 39 as well (according to the expression for R_1). Therefore, n divides 15 (according to the expression for R_2), and 9 (according to the expression for R_3), and 6 (according to the expression for R_4). Finally, n divides 3 (according to the expression for R_5). The last conclusion contradicts the assumption $n > 3$. Therefore, no number greater than 3 divides both 132 and 93, i.e., $GCD(132, 93) = 3$.

A spreadsheet is particularly amenable for such recurrent computing of the remainders. As far as the evaluation of the function $T(p)$ is concerned, by replicating the above fundamental relation among dividend, divisor, quotient, and remainder across and down a spreadsheet template (Fig. 5.5) each term of the sum

$$GCD(p,1) + GCD(p,2) + ... + GCD(p,p)$$

can be found through the Euclidean algorithm. In comparison with the environment described above (Fig. 5.4), this new computational setting has real potential to reduce the number of calculations in evaluating the function $T(p)$.

In constructing such a spreadsheet, the first meaningful inquiry is:

How many rows of a spreadsheet should one allocate in order to implement the Euclidean algorithm for a particular pair of whole numbers (k, n)?

As Fig. 5.5 shows, computing the $GCD(k,15), 0 < k < 15$, requires at most four divisions for each k. Simple calculations show, however, that the finding of $GCD(15,26)$ requires five divisions until zero remainder is reached. This raises the question about the maximum number of divisions, which may be necessary for finding the greatest common divisor of two numbers within a specified range. The answer to this question is given by a theorem due to Lamé[1]. Namely,

The number of divisions required to find the greatest common divisor of two numbers is not greater than five times the number of digits of the smaller number.

According to Lamé's theorem [Uspensky and Heaslet, 1939], in order for the Euclidean algorithm to be carried out within a spreadsheet for any two numbers, one of which has not more than two digits, ten rows of the spreadsheet should be allocated. In general, from Lamé's theorem it follows that the number of computations needed to evaluate $T(p)$ through the Euclidean algorithm equals $5 \cdot INT[\log(p) + 1]$. Here,

---

[1] Gabriel Lamé — a French mathematician of the nineteenth century.

$INT[\log(p)+1]$ is the number of digits of the number p. In comparison with the computational environment shown in Fig. 5.4 in which the evaluation of $T(200)$ requires 40,000 calculations, this new environment requires only 3000 calculations. Rather than being formally taught a piece of isolated mathematics, Lamé's theorem can be presented to teacher candidates as a tool needed to resolve a problematic situation. Acquiring such skills in mathematics pedagogy and gaining that kind of experience of mathematics with technology would help teacher candidates being creative in implementing their own technology-enabled mathematics pedagogy.

Once again, one can be engaged into problematizing the issue of data representation within a computational medium. The following questions can be explored:

How can one make a computer to display all GCDs in one line (row 17)?

How can one make a computer to terminate the execution of the Euclidean algorithm when number n reaches its upper limit displayed in cell B3?

How does one count the number of different GCDs in the sum of GCDs?

Whereas the elaboration of these questions is left to the reader, one especially engaging and relevant investigation is worthy of brief mentioning. It deals with finding a pair of whole numbers in the range not greater than 200, for which the Euclidean algorithm requires exactly ten divisions. As it turned out, such pair is 144 and 89 and the sequence of the corresponding remainders is 55, 34, 21, 13, 8, 5, 3, 2, 1. What is special about these numbers? One can recognize in this sequence the celebrated Fibonacci numbers (studied in detail in Chapter 10). Later, a teacher candidate remarked:

"In our classroom investigation of the Euclidean algorithm it was very surprising to find the appearance of the Fibonacci sequence! Why? What does it mean? Wow! The wonder of the Fibonacci numbers ... they pop up everywhere ... and what does it mean?"

The focus on meaning in this comment makes a strong case for technology-enabled mathematics pedagogy [Abramovich, 2014]. Indeed, it is the spreadsheet environment enables one to revisit Fibonacci numbers by recognizing that the relation among the dividend, quotient, divisor and remainder, $D = dQ + R$, that underlies the Euclidean algorithm, turns out to be the Fibonacci recursion, $D = d + R$, in the case of the unit quotient, when at each new step the divisor and the remainder from the previous step become, respectively, the dividend and the divisor. For example, for the pair 144 and 89, the first step of the Euclidean algorithm gives $144 = 89 + 55$ and the next step gives $89 = 55 + 34$. In particular, any two consecutive Fibonacci numbers are relatively prime. Clearly, such intriguing investigation without the use of technology may be beyond the grasp of teacher candidates. The next section provides yet another example of using the spreadsheet-based Euclidean algorithm as a tool of technology-enabled mathematics pedagogy.

		1	2	3	4	5	6	7	8	9	10	11	12	13	14	15
p																
	15															
Euclidean		1	2	3	4	5	6	7	8	9	10	11	12	13	14	15
Algorithm		15	15	15	15	15	15	15	15	15	15	15	15	15	15	15
		1	2	3	4	5	6	7	8	9	10	11	12	13	14	15
			1		3		3	1	7	6	5	4	3	2	1	15
					1				1	3		3		1		15
												1				15
																15
																15
																15
																15
																15
																15
		9	8	9	7	9	8	8	7	7	8	6	8	7	8	
GCD(C4,C5)		1	1	3	1	5	3	1	1	3	5	1	3	1	1	15
		8		4		2										1
T(*p*)	45	8		12		10										15

Fig. 5.5. Computing $T(p)$ through the Euclidean algorithm, $T(15) = 45$.

5.7.3 *Generating prime numbers through the Euclidean algorithm*
As was mentioned in the previous chapters, a spreadsheet is capable of dynamic construction of non-constructively defined functions; that is, the software is capable of creating a computerized table representation of functions defined neither through the closed-form formula nor through recursion. The function $T(p)$ is an example of such a numerical function of an integer variable p the values of which cannot be computed just by replicating a spreadsheet formula relating p to $T(p)$ or by defining $T(p)$

recursively. The use of circular references in spreadsheet formulas allows one to compute the values of $T(p)$. Figure 5.6 displays a corresponding numeric situation upon the completion of the iterative mode of computing; its bottom part (rows 23 and 24) is the table representation of the function $T(p)$, the values of p and $T(p)$ are in rows 23 and 24, respectively. Following is the description of operations that must be carried out in order to construct such table for 200 values of $T(p)$.

	A	B	C	D	E	F	G	H	I	J	K	L	M	N	O	P	Q	R	S
1			1	2	3	4	5	6	7	8	9	10	11	12	13	14	15	16	17
2		p																	
3		200																	
4	Euclidean		1	2	3	4	5	6	7	8	9	10	11	12	13	14	15	16	17
5	Algorithm		200	200	200	200	200	200	200	200	200	200	200	200	200	200	200	200	200
17	GCD(C4,C5)		1	2	1	4	5	2	1	8	1	10	1	4	1	2	5	8	1
18																			
19			80	40		20	16			20	8								
20																			
21	$T(p)$	1300	80	80		80	80			160	80								
22																			
23			1	2	3	4	5	6	7	8	9	10	11	12	13	14	15	16	17
24			1	3	5	8	9	15	13	20	21	27	21	40	25	39	45	48	33

Fig. 5.6. A fragment of the table representation of $T(p)$ displayed in rows 23 and 24.

First, as Fig. 5.6 shows, row 23 has to be filled with the values of p — consecutive natural numbers from 1 through 200, cell C24 to be entered with 1, cell D24 with the spreadsheet formula =IF($B$3=1,0,IF(D23=$B$3,$B$21,D24)), which includes a circular reference to cell D24 replicated then to cell GT24 (i.e., 199 times to the right). Second, one has to create a scroll bar that controls the range 1 through 200, link it to cell B3, set it at the starting value ($p = 1$), and play the scroll bar. In doing so, upon the appearance of every new value of $T(p)$ in cell B21, the corresponding cell in row 24 duplicates that value and leaves it unchanged in a subsequent computing due to the circular reference in the above spreadsheet formula. Note that the powerful computational feature of a spreadsheet to dynamically construct table representations of numerical functions can be utilized in other situations: generating numbers with special properties like Pythagorean triples, abundant, deficient, perfect and prime numbers; generating table representations of numerical functions like the sum of divisors function, Euler phi function (Sec. 6.8 of Chapter 6), to name a few. In particular, prior to playing the scroll bar, the spreadsheet formula =IF(D24=2*D23-

1,D23, " ") can be defined in cell D22 and replicated to the right so that the spreadsheet then becomes a generator of prime numbers.

The table representation of the function $T(p)$ can be complemented by its graphic representation (Fig. 5.7). One can recognize that the straight line $y = 2p - 1$ coincides with the graph of $T(p)$ at the points when p is a prime number (Sec. 5.7.1). Therefore, the graph of the equation $T(p) = 2p - 1$ gives an alternative representation of the distribution of prime numbers among the natural numbers. A thoughtful investigation of both the table and the graph reveals an intriguing behavior of this function and can generate a variety of follow-up activities for teacher candidates. Examples of such activities (assignments) designed to promote the learning of structured programming and data representation in a spreadsheet environment are included in Appendix.

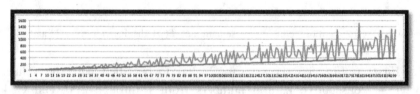

Fig. 5.7. Prime numbers as graphic solutions to the equation $T(p) = 2p - 1$.

Remark 5.3. As was mentioned above, one can avoid using the Euclidean algorithm and, instead, use the function GCD available in the spreadsheet toolkit of computing devices. This alternative allows for the generation of prime numbers in a more efficient way than through the sieve of Eratosthenes (Chapter 4). The corresponding computational environment, shown in Fig. 5.8 is programmed as follows. Cell D1 is controlled by a slider and given name p. Cell A1 is entered with 1 and cell A2 with the formula =IF(A1 < p, A1+1, " ") which is replicated down column A allowing for the value of p controlled by the slider to determine an *n*-range for which the GCD(*p*, *n*) values are computed. To this end, the formula = IF(A1= " ", " ", GCD(p, A1)) is defined in cell B1 and replicated down column B. In cell C1 (given name T) the formula =SUM(B1:B100) is defined to allow for evaluating $T(p)$ for $1 \le p \le 100$. In cell E1 the formula =IF(p=1, " ", IF(p=A1, T, E1)) with a circular reference is defined and replicated down column E. Finally, the formula =IF(OR(p=1,E1=" "), " ", IF(E1=2*A1-1, A1, " ")) is defined in cell F1 and replicated down column F. By playing the slider attached to cell D1, all prime numbers not greater than 100 can be generated. Unlike the sieve of Eratosthenes, this environment does not require using an initial set of prime numbers as a means of generating larger prime numbers.

	A	B	C	D	E	F
1	1	1	520	100		
2	2	2			3	2
3	3	1			5	3
4	4	4			8	
5	5	5			9	5
6	6	2			15	
7	7	1			13	7
8	8	4			20	
9	9	1			21	
10	10	10			27	
11	11	1			21	11
12	12	4			40	
93	93	1			305	
94	94	2			279	
95	95	5			333	
96	96	4			560	
97	97	1			193	97
98	98	2			399	
99	99	1			441	
100	100	100			520	

Fig. 5.8. Column F: solving the equation $T(p) = 2p - 1$ (rows 13-92 hidden).

Chapter 6

Pythagorean Triples

6.1 Introduction

The problem of finding integral solutions to polynomial equations in several unknowns with integral coefficients has a long history. It bears a branch of the theory of numbers known as Diophantine analysis [Steuding, 2005]. One of the most famous equations of that type is the following equation of the second degree with three unknowns

$$x^2 + y^2 = z^2 \qquad (6.1)$$

called the Pythagorean equation and named after Pythagoras — a Greek philosopher and mathematician of the sixth century B.C.[m] Any set of three positive integers satisfying Eq. (6.1) is called a Pythagorean triple. It is well known that the triple (3, 4, 5) gives a solution to Eq. (6.1). However, an example of another three positive integers satisfying Eq. (6.1) without being proportional to 3, 4, and 5 is less generally known, although a multitude of Pythagorean triples can be generated using *Wolfram Alpha* (Fig. 6.1). The triple (x, y, z) is said to be primitive if the greatest common divisor of x, y, and z is equal to one. How can one find primitive Pythagorean triples in a systematic way? The answer to this question was already known to Euclid in the form of formulas that generate right triangles with whole number sides. Although these formulas have a very simple structure, they are usually taught only in courses on number theory and are not as well known as the Pythagorean equation itself.

Consider another primitive Pythagorean triple, (5, 12, 13). How can the triple be found, neither by trial and error nor online? What is special about the number 13 that enables its square to be represented as the sum of two squares, $13^2 = 5^2 + 12^2$? In a famous discussion on

---

[m] The following quote (translated from Russian into German and then into English) can be found in Cooke [2010]: "Fifteen hundred years before the time of Pythagoras ... the Egyptians constructed right angles by so placing three pegs that a rope measured off into 3, 4, and 5 units would just reach around them, and for this purpose professional "rope fasteners" were employed" (p. 471).

Pythagorean triples within a problem-solving course, Schoenfeld [1992] acknowledged that the availability of a few triples (found by trial and error) didn't allow students to conjecture the general formulas for the triples. A quarter century after this discussion became part of the mathematics education literature, the use of *Wolfram Alpha* makes it possible to generate Pythagorean triples at the push of a button. But even the availability of a large number of triples does not allow one to guess general formulas for the triples.

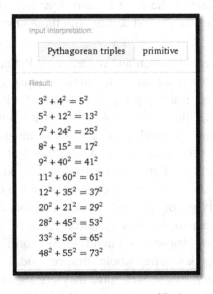

Fig. 6. 1. *Wolfram Alpha* as a generator of Pythagorean triples.

Accidentally, one can come across the identity $(n^2 + m^2)^2 = (n^2 - m^2)^2 + (2nm)^2$ which points at a way Pythagorean triples can be generated. Of course, if one knows that the formulas $z = n^2 + m^2$, $x = n^2 - m^2$, and $y = 2nm$ may be used to generate the triples, this fact can be verified numerically within a spreadsheet. Such an approach has its merit as it may be seen as developing algebraic understanding through the use of numerical evidence. However, in that case, one needs additional information about the relationship between n and m in order to generate primitive triples. An obvious assumption that n and m are relatively prime numbers turns out being too cursory. For example, when $n = 5$ and

$m = 3$, we have the triple (16, 30, 34) which, being proportional to (8, 15, 17), is not a primitive one; yet the numbers 5 and 3 are relatively prime. In what follows, it will be demonstrated how the appropriate use of a spreadsheet enhances conceptual understanding of the ancient problem, enables the discovery of underlying relationships structuring integral solutions to Eq. (6.1), and makes conjecturing of these relationships possible through a computational experiment.

6.2 Towards Finding the General Solution to the Pythagorean Equation

Modeling Eq. (6.1) within a spreadsheet is a two-dimensional computational experiment that generates data in the form of the arrays of numbers. The most important part of any such experiment is the interpretation of numeric data that is aimed at algebraic generalization. For example, one may observe that the largest element in each of the triples (3, 4, 5) and (5, 12, 13) is a prime number. This observation can be first justified through numerical evidence for several other primitive Pythagorean triples but then rejected using the triple (7, 24, 25) as a counterexample. So, not every observation is easily generalizable. This suggests looking for an alternative approach. One such alternative experiment, instead of generating a list of triples, is to generate the sums of two squared integers, $x^2 + y^2$, and then generate integers, z, whose squares can be represented as those sums to satisfy Eq. (6.1).

With this in mind, one has to take two integers, construct the sum of their squares, and then test whether this sum is a perfect square or not — a simple mathematical experiment. The major question remains: How can this be done in a systematic way? To facilitate an answer to this question, following Pólya's [1957] famous dictum, one may think about a similar problem: find two numbers whose product is a perfect square. This brings to mind the multiplication table. Indeed, in much the same way as the multiplication table can be created (Chapter 2), one can create a table of the sums of two squares. This is the first step of the experiment. The second step is to extract the square root from each of its entries and leave out the non-integer results of this operation. In doing so, Pythagorean triples can be found.

So, the alternative approach to the Pythagorean equation starts with generating the sums of two squared integers. Prior to the use of a spreadsheet, it may be helpful to chart these sums using paper and pencil (Fig. 6.2). In that way, just as in the case of the sieve of Eratosthenes (Sec. 4.2 of Chapter 4), a computational medium can be linked to a traditional, off-computer environment. Indeed, the chart pictured in Fig. 6.2 looks like a spreadsheet — a table of rows and columns.

a\\b	1	2	3	4	5
1	2	5	10	17	26
2	5	8	13	20	29
3	10	13	18	25	
4	17	20	25		
5	26	29			

Fig. 6.2. The sums of two squared integers.

So, the first step is to create the same chart using a spreadsheet (Fig. 6.3), both for the sake of accuracy and the ease of calculations. To this end, in row 1 and column A positive integers x and y, respectively, must be defined. The spreadsheet formula =$A2^2+B$1^2 must be entered into cell B2 and then copied to, say, cell K11. As always, the $ sign designates the coordinate immediately to the right to stay the same across the template. Alternatively, one can use names instead of cell references by highlighting the numbers in row 1 and column A and defining the names x and y, respectively; then the formula =x^2+y^2 can be defined in cell B2 and replicated across the spreadsheet. Either way, the range [B2:K11] becomes filled with the sums of squared integers.

At that point the following two questions can help move the experiment further:

Does the sum depend on which numbers (cells) we choose first and which numbers we choose second?

Does one need to calculate the sums that differ in the order of addends only?

	A	B	C	D	E	F	G	H	I	J	K
1		1	2	3	4	5	6	7	8	9	10
2	1	2	5	10	17	26	37	50	65	82	101
3	2	5	8	13	20	29	40	53	68	85	104
4	3	10	13	18	25	34	45	58	73	90	109
5	4	17	20	25	32	41	52	65	80	97	116
6	5	26	29	34	41	50	61	74	89	106	125
7	6	37	40	45	52	61	72	85	100	117	136
8	7	50	53	58	65	74	85	98	113	130	149
9	8	65	68	73	80	89	100	113	128	145	164
10	9	82	85	90	97	106	117	130	145	162	181
11	10	101	104	109	116	125	136	149	164	181	200

Fig. 6.3. The sums of two squares generated by a spreadsheet.

Obviously, a computer calculates each such sum twice, and a way to avoid duplications is not to do calculations in cells located below the main diagonal of the spreadsheet. In other words, to do calculations only when a number in row 1 is greater than or equal to the corresponding number in column A; that is, when $x \geq y$. It appears that the comparison of cells B1 and A2 should be used in calculations, and this implies the need to employ a logical spreadsheet function IF.

	A	B	C	D	E	F	G	H	I	J	K
1		1	2	3	4	5	6	7	8	9	10
2	1	2	5	10	17	26	37	50	65	82	101
3	2		8	13	20	29	40	53	68	85	104
4	3			18	25	34	45	58	73	90	109
5	4				32	41	52	65	80	97	116
6	5					50	61	74	89	106	125
7	6						72	85	100	117	136
8	7							98	113	130	149
9	8								128	145	164
10	9									162	181
11	10										200

Fig. 6.4. Avoiding trivial duplications of the sums of two squares.

In such a way, the further refinement of the content of cell B2 results in the formula =IF(B$1-$A2>=0,$A2^2+B$1^2," "); alternatively, using the names, =IF(x>=y,x^2+y^2," "). Now, copying cell B2 across all columns and down all rows results in the triangular array of the sums of two squared integers shown in Fig. 6.4.

Note that this array still contains same numbers: for example, the number 85 appears twice (cells H7 and J3). That is, $85 = 7^2 + 6^2$ and $85 = 9^2 + 2^2$. This, as one can see, is not a trivial duplication — the appearance of equal numbers in the spreadsheet of Fig. 6.4 (where also $65 = 8^2 + 1^2 = 7^2 + 4^2$) indicates the possibility of multiple representations of integers as a sum of two squares. As shown in Sec. 4.7 of Chapter 4, Euler used this property of integers to find the prime factorization of the fifth Fermat's "prime".

6.3 Pythagorean Triples Emerge

After the spreadsheet of Fig. 6.4 has been constructed, a new experiment that consists of testing whether a number in this array is a perfect square or not begins. The following questions can be asked:

Why is 4 a perfect square and 5 is not?

What kind of criteria can be used to distinguish between the two numbers in that sense?

These questions can help one understand that the square root of 4 is an integer, whereas the square root of 5 is not. Then, one can use the spreadsheet function INT in this new experiment. The spreadsheet employing this function is shown in Fig. 6.5 and programmed as follows. Column A and row 1 coincide with those of the spreadsheet of Fig. 6.4. The spreadsheet formula

=IF(AND(B$1>=$A2, INT(($A2^2+B$1^2)^(1/2))

=($A2^2+B$1^2)^(1/2)), ($A2^2+B$1^2)^(1/2)," "),

alternatively, using the names,

=IF(AND(x>=y, INT((x^2+y^2)^(1/2))=(x^2+y^2)^(1/2)),

(x^2+y^2)^(1/2)," "),

can be defined in cell B2 and replicated across all columns and down all rows.

The creation of two arrays of numbers shown in Figs. 6.4 and 6.5 is followed by the interpretation of the results of modeling. However, prior to the interpretation, the discussion can be accompanied by the following questions regarding the spreadsheet of Fig. 6.5:

How can Pythagorean triples be recognized in the spreadsheet?

Which triples are primitive and which are not?

How are the triples proportional to the triple (3, 4, 5) located in the spreadsheet?

What primitive triples can be observed in the spreadsheet?

How can other primitive triples be generated?

How can the triple (13, 84, 85) be located?

	A	B	C	D	E	F	G	H	I	J	K	L	M	N	O	P
1		1	2	3	4	5	6	7	8	9	10	11	12	13	14	15
2	1															
3	2															
4	3			5												
5	4															
6	5												13			
7	6							10								
8	7															
9	8															17
10	9												15			
11	10															

Fig. 6.5. Spreadsheet-generated Pythagorean triples.

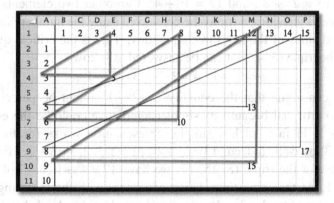

Fig. 6.6. Geometric representation of Pythagorean triples within a spreadsheet.

Furthermore, one can actually construct right triangles related to the available triples. To this end, each of the three cells can be connected with straight lines (Fig. 6.6). These questions and constructions are aimed at distinguishing between primitive triples and triples proportional to them. One can visualize how elements of the triples shape right triangles with parallel or nonparallel hypotenuses, depending on how the triples are related. The use of a spreadsheet makes it possible to integrate two different representations of Pythagorean triples — numeric and geometric — in a single drawing. Indeed, the representation of a triple in the form (3, 4, 5) does not provide any geometric information, whereas visually, the arrangement of these numbers within the spreadsheet clearly points to a right triangle.

6.4   Computer-Mediated Discourse on the Pythagorean Equation

The main goal of the ongoing activities is the discovery of a general solution to Eq. (6.1) through empirical induction; in other words, through making connections among the numbers generated by the spreadsheets of Figs. 6.5 and 6.4. The first observation that one can make by comparing the sums of two squares (Fig. 6.4) and the largest elements of Pythagorean triples (Fig. 6.5) is that some elements of the latter group can be found among the elements of the former group. Consider the numbers 5, 13, and 17 — the largest elements of the corresponding primitive triples (3, 4, 5), (5, 12, 13) and (8, 15, 17). As the sums of two squares, the three numbers have the following representations: $5 = 1^2 + 2^2$, $13 = 2^2 + 3^2$, and $17 = 1^2 + 4^2$. Can other two elements of the above three Pythagorean triples be connected using the numbers from the right-hand sides of the last three equalities? This is the critical question and for the triple (3, 4, 5) it can be answered as follows:

$$5 = 2^2 + 1^2, 4 = 2 \cdot 2 \cdot 1, 3 = 2^2 - 1^2. \tag{6.2}$$

One can argue that the equality $3 = 2^2 - 1^2$ might be perceived as being too artificial because its simpler alternative could be the equality $3 = 2 + 1$. In fact, such a simple representation works for the triple (5, 12, 13) as well. Indeed, $13 = 3^2 + 2^2, 12 = 2 \cdot 3 \cdot 2, 5 = 3 + 2$. However, in the case of the triple (8, 15, 17) we have $17 = 4^2 + 1^2, 8 = 2 \cdot 4 \cdot 1$, yet $15 \neq 4 + 1$. At the same time, being consistent with the (seemingly too artificial) equality $3 = 2^2 - 1^2$ we also have $15 = 4^2 - 1^2$. Likewise, we have $5 = 3^2 - 2^2$. Therefore, in addition to (6.2), one can express the elements of the triples (5, 12, 13) and (8, 15, 17), respectively, as follows:

$$13 = 3^2 + 2^2, 12 = 2 \cdot 3 \cdot 2, 5 = 3^2 - 2^2 \qquad (6.3)$$

and

$$17 = 4^2 + 1^2, 8 = 2 \cdot 4 \cdot 1, 15 = 4^2 - 1^2. \qquad (6.4)$$

Note that the pairs of numbers (2, 1), (3, 2), and (4, 1) involved in the right-hand sides of relations (6.2)-(6.4) are not only relatively prime but of opposite parity as well.

The patterns described through relations (6.2)-(6.4) appear to continue as one attempts testing the triples (21, 20, 29) and (55, 48, 73) for which $29 = 5^2 + 2^2$, $20 = 2 \cdot 5 \cdot 2$, $21 = 5^2 - 2^2$, and $73 = 8^2 + 3^2$, $48 = 2 \cdot 8 \cdot 3$, $55 = 8^2 - 3^2$, with the pairs (5, 2) and (8, 3) being relatively prime numbers of opposite parity. The interactive nature of the computational experiment allows one to generate new triples by simply changing the entries of cells B1 and A2 in the spreadsheet of Fig. 6.5. Alternatively, new primitive Pythagorean triples can be generated by *Wolfram Alpha* (Fig. 6.1). This can lead to new observations and discoveries through which it is quite common to touch upon profound mathematical ideas such as the multiple representations of integers as a sum of two squares.

6.5 Achieving the Goal and Going Beyond It

By generalizing from special cases and using appropriate technology-motivated revisions, one can formulate

Proposition 1. If (x, y, z) is a primitive Pythagorean triple, then one of the numbers x and y is even, and the other is odd. If y is even, then

$$x = n^2 - m^2, \quad y = 2mn, \quad z = n^2 + m^2, \qquad (6.5)$$

where m and n are relatively prime positive integers of opposite parity, $n > m$.

Proof. Formulas (6.5) can be proved by rewriting Eq. (6.1) in the form $y^2 = (z - x)(z + x)$ whence

$$z + x = y \frac{n}{m} \qquad (6.6)$$

and

$$z - x = y \frac{m}{n} \qquad (6.7)$$

where n and m are relatively prime integers of opposite parity, $n > m > 0$. Adding and subtracting relations (6.6) and (6.7) yield $2z = y(\dfrac{n}{m} + \dfrac{m}{n}) = y\dfrac{n^2 + m^2}{mn}$ and $2x = y(\dfrac{n}{m} - \dfrac{m}{n}) = y\dfrac{n^2 - m^2}{mn}$, respectively, whence

$$\frac{y}{z} = \frac{2mn}{n^2 + m^2} \tag{6.8}$$

and

$$\frac{y}{x} = \frac{2mn}{n^2 - m^2} . \tag{6.9}$$

It follows from relations (6.8) and (6.9) that any three natural numbers x, y, and z satisfying Eq. (6.1) can be expressed as follows:

$$x = r \cdot (n^2 - m^2), \; y = r \cdot 2mn, \; z = r \cdot (n^2 + m^2).$$

When n and m are of opposite parity and $r = 1$, the triple (x, y, z) is a primitive Pythagorean triple. In order to show that so defined x, y, and z satisfy the Pythagorean equation, the identity

$$[r \cdot (n^2 - m^2)]^2 + (r \cdot 2mn)^2 = [r \cdot (n^2 + m^2)]^2$$

has to be verified. Indeed, cancelling out r^2 and squaring the remaining terms yield the relation $n^4 - 2n^2m^2 + m^4 + 4n^2m^2 = n^4 + 2n^2m^2 + m^4$, a true algebraic identity. This concludes the proof.

Remark 6.1. Formulas (6.5) represent the general solution to Eq. (6.1) known to Euclid. Numbers m and n in (6.5) can be referred to as the generators of Pythagorean triples. In particular, for the pair $(n, m) = (2, 1)$ formulas (6.5) yield the triple $(3, 4, 5)$.

Remark 6.2. Although formulas (6.5) provide the general solution to equation (6.1), they are not immediately applicable to the production of relatively prime Pythagorean triples in a systematic way. To this end, one has to develop a system through which relatively prime generators n and m of different parity can be constructed. As shown in Sec. 6.8 below, this is not a simple task. One way to accomplish this task is to use the Euclidean algorithm. Another, computationally simple, way of generating primitive Pythagorean triples is through the use of formulas that can generate the triples starting from a single primitive triple, e.g., from the classic triple (3, 4, 5). There are three groups of such formulas [Weisstein, 1999, p. 1467] that can be defined recursively as follows:

$$x_n = x_{n-1} - 2y_{n-1} + 2z_{n-1},\ y_n = 2x_{n-1} - y_{n-1} + 2z_{n-1},$$
$$z_n = 2x_{n-1} - 2y_{n-1} + 3z_{n-1},\ x_1 = 3,\ y_1 = 4,\ z_1 = 5;$$

(6.10)

$$x_n = x_{n-1} + 2y_{n-1} + 2z_{n-1},\ y_n = 2x_{n-1} + y_{n-1} + 2z_{n-1},$$
$$z_n = 2x_{n-1} + 2y_{n-1} + 3z_{n-1},\ x_1 = 3,\ y_1 = 4,\ z_1 = 5;$$

(6.11)

$$x_n = -x_{n-1} + 2y_{n-1} + 2z_{n-1},\ y_n = -2x_{n-1} + y_{n-1} + 2z_{n-1},$$
$$z_n = -2x_{n-1} + 2y_{n-1} + 3z_{n-1},\ x_1 = 3,\ y_1 = 4,\ z_1 = 5.$$

(6.12)

As shown in Fig. 6.7, a spreadsheet is particularly amenable for generating Pythagorean triples using the recursive nature of Eqs. (6.10)-(6.12). Moreover, using this spreadsheet, one can test that triples generated by Eqs. (6.10)-(6.12) are primitive ones (Fig. 6.7, columns D, I, and N). Finally, *Wolfram Alpha* can be used to test the formulas in verifying Eq. (6.1) for $x = x_n,\ y = y_n,\ z = z_n$, assuming that it is true for $x = x_{n-1},\ y = y_{n-1},\ z = z_{n-1}$. Such verification in the case of formulas (6.10) is shown in Fig. 6.8. Similarly, Eqs. (6.11) and (6.12) can be put to test using *Wolfram Alpha*.

	B	C	D	E	F	G	H	I	J	K	L	M	N
1	4	5	1		3	4	5	1		3	4	5	1
2	12	13	1		21	20	29	1		15	8	17	1
3	24	25	1		119	120	169	1		35	12	37	1
4	40	41	1		697	696	985	1		63	16	65	1
5	60	61	1		4059	4060	5741	1		99	20	101	1
6	84	85	1		23661	23660	33461	1		143	24	145	1
7	112	113	1		137903	137904	195025	1		195	28	197	1
8	144	145	1		803761	803760	1136689	1		255	32	257	1
9	180	181	1		4684659	4684660	6625109	1		323	36	325	1
10	220	221	1		27304197	27304196	38613965	1		399	40	401	1
11	264	265	1		159140519	159140520	225058681	1		483	44	485	1
12	312	313	1		927538921	927538920	1311738121	1		575	48	577	1
13	364	365	1		5406093003	5406093004	7645370045	1		675	52	677	1
14	420	421	1		31509019101	31509019100	44560482149	1		783	56	785	1

Fig. 6.7. Generating Pythagorean triples using Eqs. (6.10)-(6.12).

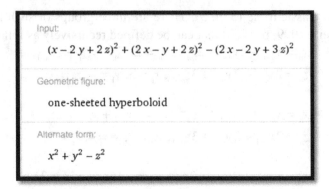

Fig. 6.8. Verifying Eqs. (6.10) using *Wolfram Alpha*.

6.6   Discovering Arithmetical Properties of Pythagorean Triangles

The spreadsheet environment enables many interesting explorations related to arithmetical properties of Pythagorean triangles. One such exploration is to locate Pythagorean triangles with equal areas. As proved by Fermat, for any integer $n > 1$, there exist n Pythagorean triangles with different hypotenuses and the same area. However, even through the use of a spreadsheet, locating Pythagorean triangles with equal areas is not an easy task. One can explore this problem for a specified area first. In other words, one can find two, three, four, *etc.*, Pythagorean triangles with different hypotenuses and same area. This does not mean that, given area, at least two Pythagorean triangles with this area can be found. Indeed, there is only one Pythagorean triangle with area 6, that is, a triangle with the side lengths 3, 4, and 5.

As an example, one can find three Pythagorean triangles with different hypotenuses each of which has area 840. To this end, one can construct a spreadsheet (Fig. 6.9) with cell A1 given the name AREA, column A and row 1 filled with positive integers (given the names x and y) and the formula

=IF(AND(x>=y,INT((x^2+y^2)^(1/2))=(x^2+y^2)^(1/2),
0.5*x*y=AREA),(x^2+y^2)^(1/2)," ")

defined in cell B2 and copied across all columns and down all rows. The logical functions IF and AND enable the spreadsheet to do multiple tests, including one that compares the area of a triangle to a number in cell A1. When one sets 840 in this cell, the spreadsheet should display a value of z in the cell defined by integers x and y (legs of a triangle), for which 840 is the half of their product. Three triples then can be found: (40, 42, 58), (24, 70, 74), and (15, 112, 113). Here we have

$$\frac{1}{2}\cdot 40 \cdot 42 = \frac{1}{2}\cdot 24 \cdot 70 = \frac{1}{2}\cdot 15 \cdot 112 = 840.$$

Note that only the third triple is a primitive one. The spreadsheet that generates the first triple is shown in Fig. 6.9.

Another simply posed problem, known as Fermat's Note XLV [Walsh, 1927-1928], is to find a Pythagorean triangle whose area is a perfect square. This investigation requires the formula

=IF(AND(x>=y,INT((x^2+y^2)^(1/2))=(x^2+y^2)^(1/2),

INT((0.5*x*y)^(1/2))=(0.5*x*y)^(1/2)), 0.5*x*y," ")

to be defined in cell B2 of a spreadsheet, in which column A and row 1 are filled with positive integers. Changing entries of cells A2 and B1, one would not be able to find such a triangle. As Fermat proved, this is indeed impossible. The two propositions about arithmetic properties of Pythagorean triangles have proof that are beyond the scope of this book. However, the questions about the properties of Pythagorean triangles — different triangles with the same area and impossibility of perfect squares as areas — are fundamental for number theory, have great historical significance, offer excellent collateral learning opportunities by making mathematical connections, and provide a rich context for learners to "interpret mathematical relationships both algebraically and geometrically" [Department for Education, 2013b, p. 42].

	A	B	C	D	E	F	G	H	I	J	K
1	840	35	36	37	38	39	40	41	42	43	44
2	35										
3	36										
4	37										
5	38										
6	39										
7	40							58			
8	41										
9	42										
10	43										
11	44										

Fig. 6.9. One of the three Pythagorean triangles with area 840.

6.7 Visualizing the Meaning of Fermat's Last Theorem

If one reads the Pythagorean equation from right to left, it can be interpreted as a representation of a square as the sum of two squares, $z^2 = x^2 + y^2$. This problem once appeared in the translation by Bachet[n] of Diophantus' *Arithmetica*. It is known that Fermat used a copy of this treatise as a notebook and left in it the intriguing message that it is impossible to write a cube as the sum of two cubes, a fourth power as the sum of two fourth powers and, in general, any power beyond the second as the sum of two like powers. For this, he wrote, there is a marvelous proof too long, however, for the margin.

This statement, made circa 1630, has become known as Fermat's Last Theorem and many of the brightest minds in mathematics have struggled to find the proof ever since. Only in 1995, due to Andrew Wiles, the proof had been found at last. As Gauss noted, "in arithmetic the most elegant theorems frequently arise experimentally as the result of a more or less unexpected stroke of good fortune, while their proofs lie so deeply embedded in the darkness that they elude all attempts and defeat the sharpest inquiries" [cited in Smith, 1959, p. 112]. Indeed, the Wiles' proof is considered enormously complex even for professional number theorists. But nowadays, it is generally accepted that the importance of a formal proof for mathematicians does not mean that the teaching of mathematics must be limited to ideas with accessible proofs [Stewart, 1990]. The use of a spreadsheet makes possible a way of visualizing the meaning of Fermat's Last Theorem by trying to find solutions to the equation

$$x^n + y^n = z^n \tag{6.13}$$

for $n > 2$ through numeric modeling in much the same way as for the Pythagorean equation ($n = 2$).

This can be done by modifying the spreadsheet of Fig. 6.4 so that the exponent n is defined in cell A1 (given the name n), the spreadsheet formula =IF(x>=y, x^n+y^n, " ") is defined in cell B2 and copied across all columns and down all rows to result in the array of the sums of two

---

[n] Claude Gaspard Bachet de Méziriac — a French mathematician of the seventeenth century.

perfect n-th powers. Then the spreadsheet of Fig. 6.5 is modified by entering the exponent n into cell A1 (given the name n) and the formula

=IF(AND(x >=y,INT((x^n+y^n)^(1/n))=(x^n+y^n)^(1/n)),

   (x^n+y^n)^(1/n)," ")

into cell B2. The latter is copied across all columns and down all rows. The logical function IF tests whether a perfect power is the sum of two like powers or not. In order to visualize the difference between the case of the second power ($n = 2$) and powers greater than two ($n > 2$), one can change the entry $n = 2$ in cell A1 for $n = 3, 4, 5$, *etc.*, and recognize the absence of integers whose cube, fourth power, fifth power, *etc.*, can be represented as the sum of two cubes, two fourth powers, two fifth powers, *etc.* This demonstrates the plausibility of the celebrated statement that Eq. (6.13) does not have integer solutions for $n > 2$.

6.8 Pythagorean Triples Meet Euler Phi Function

Can formulas (6.5) be put to work in a computational environment that generates strings of non-trivial solutions to Eq. (6.1) at the click of a button? Once again, note that this question is purely educational in the context of spreadsheets because *Wolfram Alpha* can easily generate Pythagorean triples. The environment sought has to be capable of generating the corresponding strings of relatively prime generators, n and m ($n > m$), of opposite parity. Figure 6.10 shows such an environment in which, given the value of $n = 15$, the GCD function is applied to all the integers relatively prime and of opposite parity to 15, that is, to the numbers 2, 4, 8, and 14. The environment has the single input — the value of n set in cell A1 which is given name n. In order to generate all integers relatively prime and of opposite parity to n, first note that the largest such integer is $n - 1$. Indeed, the two consecutive integers are of different parity and the assumption $GCD(n, n - 1) = k$ implies that k divides their difference which is equal to 1; that is, $k = 1$ thus indicating that two consecutive integers are relatively prime. Furthermore, the smallest number of different parity to n is either 1 (when n is even) or 2 (when n is odd). This explains the following formulas defined in cells B3, C3, and B4, respectively: =n; =IF(MOD(n,2)=0,1,2); and =IF(C3=" "," ",IF(C3=n-1," ",n)) — replicated down column B.

However, in order to continue generating values of m (greater than either 1 or 2) in column C, one has to find the GCD between n and all integers m smaller than and of different parity to n. This would allow one to select those m that are also relatively prime to n. To clarify the need to use the GCD function note that if, for example, $n = 15$, the smallest value of m is 2 and, increasing m by 2, already on the second step yields 6, an integer that is not relatively prime to 15. So, cell C4 needs a formula different from adding 2 to the previous cell. To this end, in row 2, beginning from cell G2 all integers of different parity (but not necessarily relatively prime) to n have to be defined. This can be done by defining in cell G2 the formula =$C$3 and in cell H2 the formula =IF(G2>n-2," ",G2+2) which is then replicated across row 2. The last formula stops generating candidates for m when m reaches $n - 1$. In particular, when $n = 15$, the range [G2:M2] becomes filled with all even numbers from 2 to14. Then, defining in cell G3 the formula =IF(G2=" ", " ", GCD(n, G2)) and replicating it to the right one can see that when this formula generates the number 1, the corresponding numbers in row 2 are relatively prime and of different parity to $n = 15$.

	A	B	C	D	E	F	G	H	I	J	K	L	M
1	15						2	4		8			14
2	4	n	m	x	y	z	2	4	6	8	10	12	14
3	1	15	2	221	60	229	1	1	3	1	5	3	1
4	2	15	4	209	120	241							
5	3	15	8	161	240	289							
6	4	15	14	29	420	421							

Fig. 6.10. Pythagorean triples generated through formulas (6.5).

Now, cell G1 is entered with the formula =IF(G3=1,G2," ") which is replicated to the right across row 1, thereby, yielding all integers relatively prime and of opposite parity to n. This makes it possible to enter the values of m in column C (cell C2 is given name m) beginning from cell C4 by defining in this cell the formula =IF(B4=" "," ", SMALL($G$1:$BD$1,A4)) which is then replicated down column C. In the last formula and that of immediately below, the reference $BD$1 to the upper bound of the range for the needed values of m allows for a sufficiently large n to be used in computations. Beginning from cell A3 the environment counts the number of Pythagorean triples related to the chosen value of n by counting the number of values of m obtained in the range G$1:BD$1. To this end, cell A3: =1; cell A4:

=IF(A3<COUNT(G$1:BD$1),A3+1," ") — replicated down column A; cell A2: =COUNT(A3:A100). Finally, the last set of formulas, given n in cell A1, produces all Pythagorean triples related to this value of n. Namely, cell D3: =IF(A3=" "," ",n^2-m^2) — replicated down column D; cell E3: =IF(A3=" "," ",2*n*m) — replicated down column E; cell F3: =IF(A3=" "," ",n^2+m^2) — replicated down column F. As shown in Fig. 6.10, the value $n = 15$ generates the triples (221, 60, 229), (209, 120, 241), (161, 240, 289), and (29, 420, 421) using formulas (6.5).

While this computational environment brings about systematic presentation of primitive Pythagorean triples related to a given generator n, an important educative point worth mentioning here is that the structure of the environment invites new observations and motivates new inquiries. For example, given n, how many values of m exist? In other words, how many Pythagorean triples can be generated by a particular value of n? Figure 6.10 shows that when $n = 15$ (cell A1), there are four values for m (cell A2); by changing n to 16 the spreadsheet would display the number 8 in cell A2.

By trying other values of n, one can discover that when n is even, the content of cell A2 equals the number of positive integers less than and relatively prime to n; when n is odd, the content of cell A2 is half the number of positive integers less than and relatively prime to n. In such a way, through the use of a spreadsheet for modeling Pythagorean triples, Euler phi function $\varphi(n)$ defined as the number of positive integers not greater than and relatively prime to n can be introduced in an applied context. Learning mathematics through applications, including computing applications, always motivates and gives meaning to a purely mathematical context. In doing so, one can come up with the following computationally motivated conjecture:

For a given generator n, there are $\varphi(n)$ generators m when n is even, and $\varphi(n) / 2$ generators m when n is odd.

This conjecture is another example of how meaning of a mathematical concept may stem from an inquiry through resolving which this very concept is used as a problem-solving tool [Vergnaud, 1982]. Furthermore, this conjecture, through which Pythagorean triples meet Euler phi function, shows how different mathematical concepts may be connected within a technology-enabled and application-oriented mathematical discourse.

6.9   Generalizing Pythagorean Triples via the Law of Cosines

In the rest of this chapter, several generalizations of Pythagorean triples and ways of their modeling within a spreadsheet will be discussed. Just as Pythagorean triples were motivated by geometry, these generalizations can be motivated by trigonometry. To this end, one can replace a right triangle by a scalene triangle to which the Law of Cosines, a particular case of which is the Pythagorean relation among the sides of a right triangle, can be applied. According to the Law of Cosines, which, in the context of high school mathematics, is referred to by the Common Core State Standards [2010, p. 74] as a generalization of the Pythagorean theorem, Eq. (6.1) becomes augmented by a trigonometric term $-2xy\cos\gamma$, where γ is the angle opposite to the side z.

To begin, consider the following second-degree Diophantine equation in three unknowns

$$ax^2 + bxy + cy^2 = dz^2 \qquad (6.14)$$

along with its two special cases: Eq. (6.1) and the equation

$$x^2 - xy + y^2 = z^2. \qquad (6.15)$$

In geometric terms, Eq. (6.1) describes the relationship among the sides of a right triangle; Eq. (6.15) describes the relationship among the sides x, y, and z of a scalene triangle with the side z facing $60°$ angle. Indeed, applying the Law of Cosines to such a triangle yields the equation $z^2 = x^2 + y^2 - 2xy\cos 60°$ from where, taking into account the

equality $\cos 60° = \dfrac{1}{2}$, Eq. (6.15) results. Setting in Eq. (6.14) $b = -2\cos\gamma$ and $a = c = d = 1$ results in the equation

$$x^2 - 2xy\cos\gamma + y^2 = z^2 \qquad (6.16)$$

for which Eqs. (6.1) and (6.15) are special cases. A triple (x, y, z) satisfying Eq. (6.16) will be referred to as a γ-triple. Consequently, a triple (x, y, z) satisfying Eq. (6.1) can be referred to as $90°$-triple; and a triple (x, y, z) satisfying Eq. (6.15) can be referred to as $60°$-triple. In making such connections, one can "follow the algebraic development of the cosine law and identify its relationship to the Pythagorean theorem" [Ontario Ministry of Education, 2005b, p. 51].

6.10   Technology-Motivated Transition to γ – Triples

There are many special properties among the elements of $90°$– triples worth exploring in a spreadsheet environment that are difficult, if not impossible to access without the use of technology. In particular, the number of integers with special properties among the elements of $90°$– triples that occur within a given range of generator m is fertile ground for discussion and exploration. For example, one may wonder: How many triangular numbers can be found among the elements of $90°$– triples generated via formulas (6.5) within a specific n-range? This inquiry, in turn, can serve as a springboard into the following computational experiment: If $90°$– triples were generated using formulas different from formulas (6.5), how would such a number of triangular numbers be different within the same range of a generator n if at all? At a more general level, one could inquire: Would the use of alternative formulas to generate $90°$– triples, and more generally γ – triples, open the gates to new, didactically sound, analytic and geometric spreadsheet-enabled explorations that draw on classical mathematics?

With this in mind, consider Eq. (6.16) — a trigonometric generalization of the Pythagorean equation — all integral solutions of which, according to [Buddenhagen *et al.*, 1992], are provided by the formulas

$$x = n^2 + 2mn, \; y = 2mn + 2(1 - \cos\gamma)m^2,$$
$$z = n^2 + 2(1 - \cos\gamma)(mn + m^2) \tag{6.17}$$

where m and n are relatively prime numbers. One can use *Wolfram Alpha* (Fig. 6.11, $k = \cos\gamma$) to verify that x, y, and z defined through formulas (6.17) satisfy Eq. (6.1). Alternatively, one can use *Maple* for the verification.

It follows from (6.17) that

(i) for $\gamma = 90°$: $(x, y, z) = (n^2 + 2mn, 2mn + 2m^2, n^2 + 2mn + 2m^2)$;

(ii) for $\gamma = 60°$: $(x, y, z) = (n^2 + 2mn, 2mn + m^2, n^2 + mn + m^2)$;

(iii) for $\gamma = 120°$: $(x, y, z) = (n^2 + 2mn, 2mn + 3m^2, n^2 + 3mn + 3m^2)$.

In case (i), substituting $n = u - v, m = v$ results in formulas (6.5). Indeed,

$$x = (u - v)^2 + 2(u - v)v = u^2 - 2uv + v^2 + 2uv - 2v^2 = u^2 - v^2,$$

$$y = 2(u - v)v + 2v^2 = 2uv - 2v^2 + 2v^2 = 2uv,$$

$$z = (u - v)^2 + 2(u - v)v + 2v^2 = u^2 - 2uv + v^2 + 2uv - 2v^2 + 2v^2 = u^2 + v^2.$$

Likewise, in cases (ii) and (iii), setting $n = u - v, m = v$ results, respectively, in slightly simpler formulas

$$x = u^2 - v^2, \ y = 2uv - v^2, \ z = u^2 - uv + v^2$$

and

$$x = u^2 - v^2, \ y = 2uv + v^2, \ z = u^2 + uv + v^2.$$

However, in comparison with formulas (6.5), relatively prime generators m and n in case (i) may be of the same parity and still produce primitive triples. Indeed, when m and n are both odd, formulas (6.5) always produce non-primitive triples, while this is not the case for formulas (6.17) when $\gamma = 90°$. For example, the case $(m, n) = (5, 3)$ yields the triple (55, 48, 73) which is a primitive one. Below, formulas (6.17) will be used to generate $\gamma -$ triples within a spreadsheet for an arbitrary value of γ. This not only will enable the geometrization of $\gamma -$ triples but also will open the gates to various geometric and analytic explorations of non-Pythagorean triples such as these.

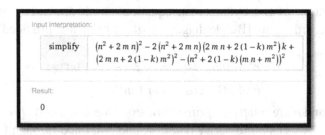

Fig. 6.11. Verifying formulas (6.17) using *Wolfram Alpha*.

The development of $\gamma -$ triples through formulas (6.17) can be interpreted geometrically as the construction of triangles with the given angle γ whose sides have integral lengths — one of the tasks that Cuoco [2001] referred to as "meta problems" (p. 171) in the context of problem posing by teachers. Apparently, integer values of the angular parameter γ do not necessarily produce rational, let alone integral, values for

γ – triples, or, geometrically speaking, the side lengths of the associated triangles. In order to insure the generation of integer γ – triples one must first restrict the values of $\cos\gamma$ to rational numbers. To address this situation, one can substitute $\dfrac{p}{q}$ for $\cos\gamma$ in formulas (6.17) and — to ensure integral triple values — can increase each of x, y, and z by the factor of q. This results in the formulas

$$x = q(n^2 + 2mn),\ y = q[2mn + 2(1 - \frac{p}{q})m^2],$$

$$z = q[n^2 + 2(1 - \frac{p}{q})(mn + m^2)]$$

(6.18)

for the construction of integer γ – triples, where p and q are relatively prime integers. Formulas (6.18) do not guarantee the triples to be primitive even if m and n as well as p and q are relatively prime pairs of integers. Indeed, the case of $n = 6$, $m = 1$, $p = 2$, and $q = 5$ yields the triple (240, 66, 222) which is not primitive. In order to generate primitive (while filtering out non-primitive) γ – triples one can use the GCD function.

	A	B	C	D	E	F	G	H	I	J	K	L	M	N	O
1															
2	cosγ =	1	2		γ=	60									
3						5									
4	9														
5	1				n	m	x	y	z	X	Y	Z	GCD(X,Y,Z)	possible m	m
6	2				9	2	117	40	103	234	80	206	2	2	2
7	3				9	4	153	88	133	306	176	266	2	3	
8	4				9	5	171	115	151	342	230	302	2	4	4
9	5				9	7	207	175	193	414	350	386	2	5	5
10					9	8	225	208	217	450	416	434	2	6	
11														7	7
12														8	8

Fig. 6.12. The generator $n = 9$ produces five 60°-triples.

As an example, Fig. 6.12 shows such a spreadsheet that generates 60°– triples. Formulas (6.18), defined in the range [I6:K6] and replicated down this range, are reduced through the application of the GCD function in column M, so that the reduction is defined in the range [F6:H6] and replicated down this range to produce primitive triples (x, y, z). Below are the programming details.

Cell A4 is slider-controlled, given name n, and it serves for defining the value of generator n. Cells B2 and C2 (given names p and q, respectively) are slider-controlled generating the values of p and q. In cell F2 the formula =ACOS(p/q)*(180/PI()) is defined and it generates the value of angle γ. Cell E6 is entered by the formula =IF(MOD(n,2)=0,1,2) which generates the smallest value of generator m relatively prime to n. In column N beginning from cell N6 all possible values of m smaller than n are defined as follows. Cell N6: = E6, cell N7 is entered by the formula =IF(N6>n-2," ", N6+1) which is replicated down column N. Consequently, in column O from the set of possible values of m only those relatively prime to n are selected through the formula =IF(N6 = " ", " ", IF(GCD(n, N6) = 1, N6, " ")) defined in cell O6 and replicated down column O.

Now, the values of m, beginning from the second smallest value, can be generated through the formula =IF(n = 1," ", IF(D7=" "," ", SMALL(O$6:O$100,A6)) defined in cell E7 and replicated down column E. Note that the reference to the upper bound O$100 of the range from which the values of m are transferred to column E can be altered to accommodate the case of a sufficiently large value of n. By entering 1 in cell A5 and the formula =IF(A5<COUNT(O$6:O$100), A5+1," ") which is replicated down column A, one can numerate generators m smaller than and relatively prime to n (in addition, the formula =MAX(A5:A100) generates in cell F3 the total number of such values of m).

The next step is to use formulas (6.18) in order to generate triples (labeled X, Y, and Z in the spreadsheet of Fig. 6.12). To this end, in cells I6, J6, and K6, respectively, the formulas

=IF(E6=" "," ",q*(D6^2+2*D6*E6)),
=IF(E6=" "," ",2*(q*D6*E6+(q-p)*E6^2)),
=IF(E6=" "," ",q*D6^2+2*(q-p)*(D6*E6+E6^2))

are defined and replicated down columns I, J, and K. Finally, in order to reduce triples (X, Y, Z) generated through formulas (6.18) to primitive triples (x, y, z), the formula =IF(I6=" "," ", GCD(I6, J6, K6)) is defined in cell M6 and replicated down column M. Now, after dividing each of the triples (X, Y, Z) by the corresponding GCD calculated in column M, primitive $60°-$triples (x, y, z) appear in columns F, G, and H. For example, one can check to see that $103^2 = 117^2 + 40^2 - 2 \cdot 117 \cdot 40 \cdot \cos 60°$

meaning that the triple (117, 40, 103) satisfies the relation among the sides x, y, and z of a triangle in which the angle 60° is opposite to the side z.

Chapter 7

Recreational Mathematics

7.1 Introduction

The problem-solving focus of current standards for school mathematics around the world [National Council of Teachers of Mathematics, 2000; Common Core State Standards, 2010; National Curriculum Board, 2008; Takahashi et al., 2006; Ministry of Education, Singapore, 2006, 2012; Ontario Ministry of Education, 2005a, 2005b; Department for Education, 2013a, 2013b, 2014] has opened the gates for recreational problems in the mathematics classroom. A characteristic didactical feature of such problems is the ease of their formulation, transparency of informal, computationally supported demonstration, and direct appeal to the natural curiosity of the 21st century students "who are digital natives comfortable with the use of technologies" [Ministry of Education, Singapore, 2012, p. 2]. Recreational problems, for instance, may include analyzing "puzzles and games that involve numerical reasoning using problem-solving strategies" [Western and Northern Canadian Protocol, 2008, p. 33]. However, the outer simplicity of recreational mathematics often turns into intrinsic complexity of formal demonstration with means and machinery far beyond the reach of even mathematically motivated students. In many such cases, a transparent mathematical statement associated with a recreational problem remains a conjecture not only for the students but for professional mathematicians as well [Williams, 2002]. A commonly accepted didactic approach to informal demonstration of such conjectures is by exploring special cases. Using a spreadsheet has great potential to mediate presentations of that kind through a numerical approach.

One class of problems in recreational mathematics deals with the properties of whole numbers formulated in terms of their digits. In this regard, two intuitively understandable concepts related to a whole number and comprising "specific topics for which understanding is critical" [National Curriculum Board, 2008, p. 4] may be singled out — a digit (face value) and its positional rank (place value). Human ability to visually identify positional rank of a digit in a whole number is essentially intuitive and the possession of such ability does not imply that one has conceptual understanding of the meaning of each digit. Yet, a spreadsheet does not have such intuitive ability and if one attempts to 'teach' the software to do what a human does automatically, one has to conceptualize the meaning of a particular digit (face value) in a number.

For example, an intuitive method of identifying four as the second digit of the number 243 is based on one's ability to visually distinguish the order (positional rank) of a digit within a whole number. Communicating this method to a spreadsheet, or replacing a visual strategy by a computational procedure, requires a conceptual understanding and mathematically formulated meaning of what is called 'the second digit'. A formal procedure is needed that can mediate the counting of digits and enables the separation of a digit from a number. As will be shown below, such a procedure is based on the meaning of a digit in the number that refines the notion of face value and stems from the conceptualization of division as measurement. The development of this procedure in the context of mathematical preparation of teachers beyond the elementary level is in agreement with the Conference Board of the Mathematical Sciences [2012] recommendations for the teachers that include the importance of learning how "to build computational models of mathematical objects … [thus avoiding using technology] in a superficial way, without connection to mathematical reasoning, [and thereby] without advancing learning" (p. 57).

7.2   A Real-Life Context for Recreational Mathematics

The measurement model for division enables for the construction of spreadsheet-based computational environments that can be used in exploring a variety of intriguing properties of integers formulated in terms of their digits. One such a property is for a number to be a palindrome, that is, to be read forward the same as backward. As suggested in one of the educational documents of the State of New York [New York State Education Department, 1996], the context in which the author prepares teachers, investigating number patterns through palindromes is an appropriate activity for students to "appreciate the true beauty of mathematics, and construct generalizations that describe patterns simply and efficiently" (p. 21). In agreement with the approach taken in this book, when context motivates the use of a spreadsheet in problem solving, palindromes can be introduced through the following

Odometer Challenge: *Mr. Jones and his eleven-year-old daughter Kelly were traveling by car in the state of New York at a constant speed from Massena to Niagara Falls. In Potsdam they stopped at a traffic light and Kelly noticed the odometer reading. "Dad, look at the odometer. It shows 16961. What an interesting number, it reads the same forward as it does backward."*

Mr. Jones looked at the odometer and confirmed Kelly's observation. "Good observation Kelly", Mr. Jones said. "You don't see a number like that often".

Kelly thought for a moment, then proposed the following to her father: "How about every time such a number appears on the odometer, that is, a number that reads the same forward as backward, I get $100 to spend on our vacation".

"No problem!" said Mr. Jones who believed such a number would not appear for a very long time.

Two hours later, when they stopped at another traffic light, the odometer once again showed a number reading the same forward as backward. Kelly made her father aware of this and he was surprised to see that a number appeared again so quickly. Kelly smiled knowingly. "Dad," she said, "I know we would soon see another number like this in about two hours. These numbers are called palindroms, we learned about them in our math class this year".

There exist many questions that turn this (recreational) real-life situation into a mathematical investigation, among them:

What is the new reading on the odometer showing a palindrome?
When will the third palindrome appear?
Is there a pattern in the sequence of palindromic numbers?
At what speed was the car traveling?
How much money would Kelly collect by the time they reach Niagara Falls if the distance from Potsdam to Niagara Falls is 310 miles?

In what follows, these questions will be answered by using a spreadsheet programmed to recognize a palindrome by separating digits from a number and then comparing the digits equidistant from the beginning and the end of the number. In order to construct such a spreadsheet, new computational tools should be developed. Mathematically, these tools stem from the development of a concept discussed in the next section.

Remark 7.1. Although *Wolfram Alpha* identifies 16961 as a palindrome in response to the query "is 16961 a palindrome number?", the program does not understand a query seeking the next palindrome number. This observation justifies the subsequent use of a spreadsheet in the computational resolution of the Odometer Challenge. Such use of a spreadsheet may give an idea of how *Wolfram Alpha* was taught to recognize palindromes.

7.3   Measurement Model for Division as a Tool in Computing Applications

Consider the following problem:

Andy has 86 marbles. He puts them in boxes containing 10 marbles each. How many boxes does Andy need?

One way to solve this problem is to subtract 10 from 86 repeatedly as many times as possible within a set of positive numbers. The meaning of the digit 8 in the number 86 is the measure of 86 by the sets of tens. In other words, dividing 10 into 86 determines how many sets of 10 marbles can be created out of 86 marbles. The digit 6, in turn, is the measure of remainder in this process. Conceptualization of division as repeated subtraction is what is commonly referred to as the measurement model for division. The problem with marbles can be used to build upon one's familiarity with this model in the context of teaching/learning division at the elementary school level and extend it to computing applications dealing with the properties of whole numbers formulated in terms of their digits. As mentioned by the Advisory Committee on Mathematics Education [2011], "developing a sound understanding of key mathematical ideas ... requires extensive experience of working with the basic mathematical concepts that underpin its range of applications"

(p. 2). In particular, the measurement model for division enables for the construction of spreadsheet-based computational environments that can be used in exploring the properties of palindromes.

7.4 A Numeric Example

Consider a base-ten number 243. This number can be represented as the sum of products of its face values (digits) and the corresponding place values (powers of ten):

$$243 = 2 \cdot 10^2 + 4 \cdot 10^1 + 3 \cdot 10^0 \qquad (7.1)$$

Relation (7.1) is a representation of the number 243 in the expanded notation as a trinomial in the powers of ten. Fig. 7.1 shows an alternative, iconic representation (using base ten blocks) of the number 243 that consists of two flats (hundreds), four longs (tens), and three units (ones). Grouping elements into the sets of hundreds (division by 100) results in a set of two flats. Grouping elements into the sets of tens (division by ten) requires the transformation of flats into longs and then removing all longs included into blocks (removing ten times the number of flats from the total number of longs). Grouping elements into the sets of ones requires the transformation of flats into longs, longs into blocks, and then removing the number of blocks included into all the longs. For example, four longs (alternatively, the second digit 4) represent the total number of such longs (24) minus ten times the number of flats (because each flat contains ten longs); that is, $4 = 24 - 10 \cdot 2$.

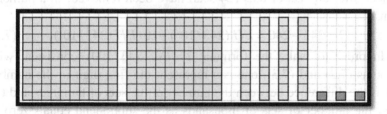

Fig. 7.1. Iconic representation of the number 243.

The iconic representation of the number 243 (Fig. 7.1) may clarify the meaning of each digit in terms of the measurement model for division. What is the meaning of the first digit, 2, in relation (7.1)? Clearly, the digit 2 shows the number of hundreds in this number. Within a physical environment, in order to determine the number of hundreds in a set, one has to group the elements into the sets of hundreds. In the numeric domain, a formal operation that defines such grouping stems from the measurement model for division. More specifically, dividing 243 by 100 and leaving out the remainder enable, with the help of the greatest integer function INT(x) returning the largest number smaller than or equal to x, the identification of the first digit (positional rank one) of the given number. In such a way, $2 = INT(243/100)$; that is, the operation shown in the right-hand side of the last equality gives the number of hundreds in the number 243.

What is the meaning of the second digit, 4, in the number 243? It represents the remaining number of tens in this number after all sets of hundreds have been removed. On an action level, in order to determine the number of tens in a set, one has to group its elements into the sets of tens. A formal operation that defines such grouping is the division by ten. However, the original number includes tens that go into hundreds also (see flats in the manipulative model of Fig. 7.1). In order to find out how many tens remain after removing all sets of hundreds, one has to divide 10 into 243, apply the greatest integer function to the quotient, and subtract ten times the number of hundreds. Numerically, this can be expressed as follows:

$$4 = 24 - 20 = INT(243/10^1) - 10 \cdot INT(243/10^2). \qquad (7.2)$$

In much the same way, the last (third) digit, 3, in the number 243 can be defined as the total number of sets of ones that can be made out of the number after all sets of tens (24 sets) have been removed. In a numeric form,

$$3 = 243 - 240 = INT(243/10^0) - 10 \cdot INT(243/10^1). \qquad (7.3)$$

In order to make the formula $2 = INT(243/100)$ consistent with formulas (7.2) and (7.3), one can represent the first digit in the number 243 as the difference between the number of the sets of hundreds and ten times the number of sets of thousands as the subtrahend equals zero in the case of a three-digit number. That is,

$$2 = 2 - 0 = INT(243/10^2) - 10 \cdot INT(243/10^3). \qquad (7.4)$$

Relations (7.2)-(7.4) have similar structure and they depend on the exponents in the powers of 10 (and, ultimately, on the digits' positional ranks) while all other components remain constant across the relations. In such a way, the notion of a face value can be conceptualized through a computational procedure (algorithm) that depends on a digit's positional rank. By using relations (7.2)-(7.4), the sum of digits of the number 243, $SD(243)$, can be found as follows

$$SD(243) = INT(243/10^2) + INT(243/10^1) - 10 \cdot INT(243/10^2)$$

$$+243 - 10 \cdot INT(243/10^1) = 243 - 9 \cdot [INT(243/10^2) + INT(243/10^1)]$$

$$= 243 - 9 \cdot (2 + 24) = 243 - 234 = 9.$$

Indeed, $2 + 4 + 3 = 9$.

7.5 The Case of an n-Digit Number

Let an n-digit integer N in the decimal positional system have the following representation

$$N = a_{n-1} \cdot 10^{n-1} + a_{n-2} \cdot 10^{n-2} + ... + a_{n-i} \cdot 10^{n-i} + a_{n-i-1} \cdot 10^{n-i-1} + ... + a_1 \cdot 10 + a_0,$$

where $0 \le a_i < 10$ for $0 \le i < n-1$ and the leading digit $1 \le a_{n-1} < 10$. Then its digits a_i have the following representation through the function $INT(x)$:

$$a_i = INT(\frac{N}{10^i}) - 10 \cdot INT(\frac{N}{10^{i+1}}), 0 \le i \le n-1. \qquad (7.5)$$

Indeed,

$$INT(\frac{N}{10^i}) - 10 \cdot INT(\frac{N}{10^{i+1}})$$

$$= INT(\frac{a_{n-1} \cdot 10^{n-1} + a_{n-2} \cdot 10^{n-2} + ... + a_i \cdot 10^i + a_{i-1} \cdot 10^{i-1} + ... + a_0}{10^i})$$

$$-10 \cdot INT(\frac{a_{n-1} \cdot 10^{n-1} + a_{n-2} \cdot 10^{n-2} + ... + a_{i+1} \cdot 10^{i+1} + a_i \cdot 10^i + ... + a_0}{10^{i+1}})$$

$$= a_{n-1} \cdot 10^{n-i-1} + a_{n-2} \cdot 10^{n-i-2} + ... + a_{i+1} \cdot 10 + a_i$$

$$-(a_{n-1} \cdot 10^{n-i-1} + a_{n-2} \cdot 10^{n-i-2} ... + a_{i+1} \cdot 10) = a_i.$$

A similar formula returns

$$d_i = INT(\frac{N}{B^i}) - B \cdot INT(\frac{N}{B^{i+1}}), 0 \le i < m, \qquad (7.6)$$

the i-th digit d_i of the same number, N, in base B, where $m = INT(\log_B N + 1)$ — the number of digits of N in base B.

Indeed, setting

$$N = d_m \cdot B^m + d_{m-1} \cdot B^{m-1} + \ldots + d_{i+1} \cdot B^{i+1} + d_i \cdot B^i + \ldots + d_1 \cdot B + d_0,$$

as the representation of N in base B through the digits d_i, it follows that

$$INT(\frac{N}{B^i}) - 10 \cdot INT(\frac{N}{B^{i+1}})$$

$$= INT(\frac{d_m \cdot B^m + d_{m-1} \cdot B^{m-1} + \ldots + d_i \cdot B^i + d_{i-1} \cdot B^{i-1} + \ldots + d_0}{B^i})$$

$$-10 \cdot INT(\frac{d_m \cdot B^m + d_{m-1} \cdot B^{m-1} + \ldots + d_{i+1} \cdot B^{i+1} + d_i \cdot B^i + \ldots + d_0}{B^{i+1}})$$

$$= d_m \cdot B^{m-i} + d_{m-1} \cdot B^{m-i-1} + \ldots + d_{i+1} \cdot B + d_i$$

$$-(d_m \cdot B^{m-i} + d_{m-1} \cdot B^{m-i-1} + \ldots + d_{i+1} \cdot B) = d_i.$$

Finally, formulas (7.5) and (7.6) lead to the following formulas for the sum of digits of the number N calculated in bases 10 and B, respectively:

$$a_0 + a_1 + a_2 + \ldots + a_n$$

$$= N - 9 \cdot [INT(N / 10) + INT(N / 10^2) + \ldots + INT(N / 10^n)] \qquad (7.7)$$

and

$$d_0 + d_1 + d_2 + \ldots + d_m$$

$$= N - (B - 1) \cdot [INT(N / B) + INT(N / B^2) + \ldots + INT(N / B^m)]. \qquad (7.8)$$

For example, a base ten number 243 ($N = 234$) can be written as a trinomial in the powers of eight

$$243 = 3 \cdot 8^2 + 6 \cdot 8^1 + 3 \cdot 8^0$$

where the coefficients 3, 6, and 3 are the digits of 243 in base eight ($B = 8$). We have $3 + 6 + 3 = 12$; according to formula (7.8),

$$3 + 6 + 3 = 243 - (8 - 1)[INT(243 / 8) + INT(243 / 8^2)]$$

$$= 243 - 7(30 + 3) = 243 - 231 = 12.$$

Likewise, when $B = 5$ we have

$$243 = 1 \cdot 5^3 + 4 \cdot 5^2 + 3 \cdot 5^1 + 3 \cdot 5^0$$

and therefore, $1 + 4 + 3 + 3 = 11$. According to formula (7.8),

$$243 - (5 - 1)[INT(243/5) + INT(243/5^2) + INT(243/5^3)]$$
$$= 243 - 4(48 + 9 + 1) = 11.$$

Formulas (7.7) and (7.8) can be used for the development of spreadsheet-based solutions to recreational problems, some of which are presented as ideas for individual projects in Appendix.

To conclude this section, note that the importance of teaching multiple base systems was theoretically enunciated by Vygotsky [1962] who argued, "The ability to shift at will one system to another (e.g., to "translate" from the decimal system into one that is based on five) ... indicates the existence of a general concept of a system of numeration" (p. 115). More recently, the Conference Board of the Mathematical Sciences [2001] emphasized the need for prospective teachers to appreciate "the place value structure of our number system, which implicitly expresses numbers as polynomials in powers of 10 and permits single-digit arithmetic to be easily extended to multi-digit arithmetic" (p. 5). A similar structure, as was shown above, can be observed in the case of bases different from ten. In the practice of mathematics teacher education, the following reflection by a teacher candidate on the author's course is worth mentioning: *"When children start doing base-ten arithmetic, it must be as confusing to them as other base systems to me. If I could develop a conceptual understanding of arithmetic in other bases, this technique should also be helpful in teaching students base ten for the first time. When students don't understand the concept and what is actually going on, math seems mysterious and more confusing. As teachers, it's important that we use techniques to help students really understand the concept behind what they are doing. By doing this, it will help make students more successful"*.

7.6 Computational Separation of Digits Within a Spreadsheet

One can construct a spreadsheet for identifying and separating digits according to their positional rank in a whole number. Formula (7.5) has an explicit reference to a positional rank enabling for the replacement of symbolic definition by the ostensive definition (Sec. 1.2 of Chapter 1) which consists of pointing by a cursor to a cell of a spreadsheet. Once displayed in a single cell, the result of this ostensive definition (i.e., a spreadsheet formula) can be replicated to other cells. When replicating such a single cell down or to the right in a chart one might overlook the fact that the length of the replication depends on the number of digits involved — a factor that may vary from number to number. A visual strategy that one uses in deciding the termination of the fill down/right action is fundamentally automatic. The need for the replacement of an automatic decision by a rigorous mathematical algorithm brings about a new problematic situation of the computational environment and, consequently, it creates a context for introducing a new mathematical concept.

	A	B
1	Positional rank	12321
2	1	1
3	2	2
4	3	3
5	4	2
6	5	1
7	6	
8	7	

Fig. 7.2. Computational separation of digits.

The expanded representation of an n-digit number prompts the conceptualization of the meaning of the largest exponent of its base. With this in mind, the concepts of logarithmic function and composition of functions can be introduced in this context from an applied perspective. Clearly, this perspective highlights meanings behind abstract concepts and ultimately leads to a better appreciation of mathematical ideas by teacher candidates.

More specifically, in order to determine the number of digits of a whole number computationally, the composition of the greatest integer function *INT* and the base ten logarithmic function *LOG* will be used below. For example,

$$INT(LOG(243)+1) = INT(2.39...+1) = 3$$

is the number of digits in the number 243. Although both functions belong to the tool kit of available computing devices of the spreadsheet environment, its syntactic versatility allows for the function $INT(LOG(x)+1)$ to be replaced by a single spreadsheet function $LEN(x)$, which returns the number of characters in a text string x.

Therefore, when applied to a whole number, the latter function returns the number of digits in this number. The recognition of the availability of different syntactic means that a spreadsheet environment can provide, contributes to a classroom culture in which flexibility and freedom in developing different problem-solving strategies constitute an important component of the environment. In such an environment "students will develop habits and strategies that will help them be better and more independent learners" [Ministry of Education, Singapore, 2012, p. 31].

Figure 7.3 shows how the use of the spreadsheet formula

=IF(A2<=LEN(B$1), INT(B$1/10^(LEN(B$1)-A2))-

10*INT(B$1/10^(LEN(B$1)-A2+1)), " "),

defined in cell B2 and replicated down column B, enables the computational separation of digits in a number displayed in cell B1. When a positional rank (displayed in column A) is not greater than the total number of digits in the number defined as LEN(B$1), this formula instructs the computer to fill a cell with the corresponding digit (according to definition (7.5)) or, otherwise, leave the cell blank.

	A	B	C
1			
2	Positional rank	2	New base
3	in the new base	9	Base 10 number
4	1	1	
5	2	0	
6	3	0	
7	4	1	
8	5		

Fig. 7.3. Computational separation of digits of 9 in base 2.

A similar spreadsheet formula (based on formula (7.6)) can be used to computationally transform a base ten number into a string of digits that represents the number in a different base. For example, $9 = 1 \cdot 2^3 + 0 \cdot 2^2 + 0 \cdot 2^1 + 1 \cdot 2^0$, that is, the (base 10) number 9 in base 2 is the string 1001. The following formula defined in cell B4 (Fig. 7.3) and replicated down column B enables the separation of digits in this (or any other) string:

=IF(INT(LOG(B$3)/LOG(B$2))+1-A4>=0,

INT(B$3/B$2^(INT(LOG(B$3)/LOG(B$2))+1-A4))

-B$2*INT(B$3/B$2^(INT(LOG(B$3)/LOG(B$2))+1-A4+1))," ") .

Here B$3 is a given number in base 10, and B$2 is a new base.

The difference between the two spreadsheet formulas is that the latter uses INT(LOG(...)/LOG(B)+1) — a spreadsheet translation of the function INT(LOG$_B$(...)+1) — in place of LEN. For example, the number of digits in the string 1001 that represents 9 in base 2 results from the equality INT(LOG(9)/LOG(2)+1) = 4. Through this approach, a concept of logarithm can become a problem-solving tool of the computational environment.

Remark 7.2. *Wolfram Alpha* is capable of finding the sum of digits of a whole number in base ten as well as in other bases. The program can convert any base ten number into a different base. Also, specific digits can be found step by step by using formula (7.5) and replacing the INT notation by that of FLOOR (Fig. 7.4).

Fig. 7.4. Using formula (7.5) and *Wolfram Alpha* in finding the second digit of 243.

7.7   Illustration 1: Resolving the Odometer Challenge

Although recognizing a palindrome visually may be easier than computationally, a spreadsheet has a clear advantage over a human when browsing through a large set of numbers in search of a palindrome. To do such a search, a spreadsheet can be used for deciding whether a given number (or a string of digits) is a palindrome or not.

More specifically, in order to recognize a palindrome one has to compare digits that are equidistant from the beginning and the end of the string (Fig. 7.5). Numerically, these digits can be paired via mediational referents (consecutive natural numbers starting from one) displayed in column A: the sum of two numbers in each pair is one greater than the number of digits in the string. To this end, one can use the spreadsheet function INDEX as a tool for identifying a palindrome. This function enables a computer to choose any number within the one-dimensional array displayed in column B given its position in this array. For example, the function =INDEX(B$3:B$20,LEN(B$2)-A3+1,1) instructs the software to choose a number within the range B$2:B$20 which position (row) is defined by a variable while the third part of the function indicates that the array comprise one column only. In such a way, one can define in cell C3 the following formula:

=IF(A3>LEN(B$2),0,IF(B3=INDEX(B$3:B$20,
LEN(B$2)-A3+1,1),1,0))

and replicate it down column C.

	A	B	C
1		NUMBER	
2	Positional rank	12321	PALINDROME
3	1	1	1
4	2	2	1
5	3	3	1
6	4	2	1
7	5	1	1
8	6		0

Fig. 7.5. Computational identification of a palindrome.

The use of a nested function IF in the last formula enables the spreadsheet to fill a cell with the number 0 when either the corresponding positional rank in column A becomes greater than the number of digits in the tested number or the comparison of the digits in column B does not account for a palindrome; otherwise, it fills the cell with the number 1 (the case opposite to the above two requirements). So in the case of a palindrome the number of cells filled with 1's should be equal to the number of digits in the tested number. The use of the formula

=IF(SUM(C3:C22)=LEN(B3),"PALINDROME!"," ") defined in cell C2 allows for such identification of a palindrome.

	A	B	C	D	E	F	G
1		NUMBER			NUMBER		DIFFERENCE
2	Positional rank	16961	PALINDROME		17071	PALINDROME	110
3	1	1	1		1	1	
4	2	6	1		7	1	
5	3	9	1		0	1	
6	4	6	1		7	1	
7	5	1	1		1	1	
8	6		0			0	

Fig. 7.6. Resolving the Odometer Challenge.

In order to resolve the *Odometer Challenge*, one can construct a similar environment for the identification of a palindrome when its search is carried out beginning from the number 16961 as the initial mileage. By changing this number in cell E2 using a slider shown at the top of column F (Fig. 7.6), one can locate the next palindrome, 17071, which differs from the palindrome 16961 by 110 (cell G2).

The spreadsheet shown in Fig. 7.6 allows for the resolution of the *Odometer Challenge*. The environment includes two consecutive identifiers of a palindrome where the left part shows 16961 as a palindrome and the right part, using a slider attached to cell E2, allows for changing its content beginning from 16962 until the next palindrome, 17071, appears. The cell G2 calculated the difference, 110, between the two palindromes, thereby indicating that when travelling 110 miles over two hours at a constant speed, the speed is 55 m/h. Likewise, the smallest palindrome greater that 17071 is 17171, with the difference 100 between the palindromes. In that way, Kelly would get $100 when the odometer shows 17071 and another $100 when the odometer shows 17171. The next palindrome is the number 17271 — the reading of the odometer when the car arrives to Niagara Falls thus travelling 110 + 100 + 100 = 310 miles — the distance from Potsdam to Niagara Falls in the state of New York. Similar problems can be formulated in other geographic and/or recreational contexts keeping in mind that an important teaching strategy "of making the mathematics curriculum accessible is to emphasise the relevance of the content to students" [National Curriculum Board, 2008, p. 6].

7.8 Illustration 2: Palindromic Number Conjecture

Palindromes have an interesting property to attract whole numbers under the following simple procedure. Start with any whole number, reverse its digits and add the two numbers. Repeat the process with the sum (new number). Continue the process to see that it leads to a palindrome. Indeed, after the verification of many special cases the following simple statement appears to be a plausible generalization:

Regardless of a starting number, one can always arrive at a palindrome by adding two numbers with digits reversed.

A spreadsheet-based investigation of this *Palindromic Number Conjecture* [Weisstein, 1999], mentioned by Williams [2002] among 12 unsolved problems in contemporary mathematics, can serve as an illustration of using technology as a cultural amplifier of discovery-oriented mathematical activities. For example, when the starting number is 39, the following two-step procedure leads to a palindrome: 39 + 93 = 132 and 132 + 231 = 363 — a palindrome. Likewise, 58 + 85 = 145 and 143 + 341 = 484 — a palindrome.

In many cases, however, the process may take more than two steps and a computational approach is useful in demonstrating this interesting property of integers through exploring the conjecture. It should nonetheless be noted that according to Gruenberger [1984], the number 196, being the smallest number of that kind, does not yield a palindrome even after 50,000 steps and, moreover, "among the first 100,000 integers there are 5,996 that apparently do not generate a palindrome no matter how long the procedure is continued (although this conjecture has not been confirmed)" (p. 24).

Finally, extending the environment shown in Figs. 7.4 and 7.5, one can define in cells D2 and C1 (Fig. 7.7) the formulas
=B1+B5+B6*10+B7*10^2+B8*10^3+B9*10^4+B10*10^5+B11*10^6+
B12*10^7+B13*10^8+B14*10^9+B15*10^10+B16*10^11+B17*10^12
+B18*10^13+B19*10^14

and
=IF(SUM(C6:C21)=INT((LEN(B$2)+1)/2),"PALINDROME"," "),
respectively, and replicating the block of cells (C1:D3) to the right,
arrive at an environment for the demonstration of the Palindromic
Number Conjecture (Fig. 7.7). Note that the length of the formula
defined in cell D1 results from the limitation of a spreadsheet — the
program does not store numbers with more than 15 digits to guarantee
precision in calculations.

Fig. 7.7. For 461 (cell B2) it requires three steps (cell A3) to reach 2662 (cell H2).

7.9 Converging Differences

Another digit-related recreational problem for which a spreadsheet can
be a useful problem-solving tool is very similar to the Palindromic
Number Conjecture. There are numbers that under certain operations
on digits can reproduce themselves over and over. Furthermore, these
numbers have a property to attract other numbers under the same
operation on digits. For example, think of a three-digit number in
which digits are not all the same. Arrange the digits in the decreasing
order; that is, write down the largest possible number using these three
digits. Let so constructed number have the form $100a + 10b + c, a > c$.
Then, write down the smallest possible number using the same digits.
Subtract the smaller number from the larger number — this operation
creates the greatest difference out of the digits involved. Because
$100a + 10b + c - 100c - 10b - a = 99(a - c)$, the difference is always a
positive multiple of 99. When the difference is equal to 99, one may
consider 99 as an artificial three-digit number 099 which becomes 990
after arranging its three digits in the decreasing order. With this new
number repeat the same procedure and continue it.

For all three-digit numbers this procedure, as it has been discovered
first by Kaprekar [1949], converges to 495. This number has a unique
property of being an attractor for all three-digit numbers through the

procedure of creating and transforming the greatest differences out of three (not all the same) digits [Trigg, 1974]. Indeed, the equality

$$100a + 10b + c - (100c + 10b + a) = 100c + 10a + b$$

(which describes the case when the greatest difference, $100c + 10a + b$, and two numbers that created it, $100a + 10b + c$ and $100c + 10b + a$, all constructed of the same digits) is equivalent to $89a - 199c = b$. This equality defines the self-reproducing endpoint because making the largest difference out of the number $100a + 10b + c$ yields the same number. The spreadsheet of Fig. 7.8 shows that the only possible triple which satisfies the latter relation is $a = 9, b = 5, c = 4$. This proves computationally that $100c + 10a + b = 495$ and this number represents a self-reproducing endpoint for all three-digit numbers. The spreadsheet formula

=IF(INT((89*B$1-199*$A2)/10)=0, 89*B$1-199*$A2," ")

that allows for the search of the number 495, is defined in cell B2 and replicated to cell J11 (Fig. 7.8). This formula, using a three-dimensional modeling technique (Chapter 3), verifies whether the difference $89a - 199c$ is a single digit number and, if so, it generates this number so that the remaining two digits a and c can be revealed as well.

	A	B	C	D	E	F	G	H	I	J
1		1	2	3	4	5	6	7	8	9
2	0									
3	1									
4	2									
5	3									
6	4							5		
7	5									
8	6									
9	7									
10	8									
11	9									

Fig. 7.8. Computational search of a self-reproducing endpoint.

7.10 Exploring the Growth of the Number of Digits in Integer Exponents

This section will investigate the growth of the number of digits in integer exponents using a spreadsheet. Through this investigation, interesting patterns can be discovered. However, it will also be shown that while some patterns in the sequences of numbers can be associated with sequences found in the On-line Encyclopedia of Integer Sequences® (OEIS®), they break down and are not confirmed for the large exponents.

It is well known that in base ten, the number of digits in the consecutive powers of ten is represented by the sequence of consecutive natural numbers starting from one. Indeed, in the sequence 10^0, 10^1, 10^2, 10^3, 10^4, … the number of digits is given by the corresponding exponents increased by one: 1, 2, 3, 4, 5, … . The same is true for the sequence B^0, B^1, B^2, B^3, B^4, … in base B. But how does the sequence of the number of digits in the last sequence looks in base 10 when B is different from ten? For example, when $B = 6$ the first four terms representing the number of digits are 1, 1, 2, 3, 4. As mentioned in Sec. 6 of this chapter, in base ten the number of digits of B^n is equal to $INT(n \cdot LOG_{10}(B)+1)$. The use of logarithms makes it possible to scale large numbers to their logarithms and therefore to avoid the limitation of a spreadsheet to store with precision only numbers with at most 15 digits.

Consider the spreadsheet pictured in Fig. 7.9. In column A the sequence of natural numbers (starting from zero) is defined. In a slider-controlled cell B2 the base of an exponent is defined and is given the name Base. In cells B3, C3, D3, E3, and F3 the following five formulas are defined, respectively,

=INT(A3*LOG10(Base)+1),
=IF(A5>MAX(B$3:B$1001), " ",COUNTIF(B$3:B$1001,A5)),
=IF(C3=" "," ",IF(ABS(C3-D$2)>0,A3," ")),
=IF(A3>COUNT(D$3:D$1001), " ", SMALL(D$3:D$1001,A4)),
=IF(E4=" ", " ", E4-E3),

and replicated down respective columns. The first formula calculates the number of digits in the corresponding integer exponent; the second formula calculates the number of appearances of the corresponding number of digits displayed in column B; the third formula seeks for the number of appearances being greater than the most common number and displays the corresponding exponent; the fourth formula collects all the exponents displayed in column D and put them as a sequence in column E; the fifth formula attempts to discover a regularity in the appearances of those exponents.

For example, in Fig. 7.9 an interesting pattern in the growth of the number of digits of the sequence $x_n = 6^n$ as the function of n has been displayed. The exponents for which the corresponding powers twice have the same number of digits within the same place value are given by the sequence 1, 4, 8, 11, 15, 18, … . Using the OEIS®, one can find out that this sequence can be expressed through the following first order recurrence equation $x(n) = 7n - x(n-1) - 9, n \geq 1, x(1) = -3$, and it represents numbers congruent to $\{1, 4\}$ modulo 7. However, generating this sequence within a spreadsheet for sufficiently large values of n shows that the pattern breaks down for $n = 68$; nonetheless, after that every second number of each sequence, $x(n)$ and the one generated by the spreadsheet in column E, appears to be the same. This example shows that purely numeric approach to a mathematical exploration that seeks pattern recognition has serious deficiency in the absence of rigorous reasoning. One does need rigor to explain why a certain pattern works.

	A	B	C	D	E	F
1	expons	Base	appearances of the number of digits	mode	outsider expons	search for a pattern
2		6		1		
3	0	i	0	0	0	
4	1	1	2	1	1	1
5	2	2	1		4	3
6	3	3	1		8	4
7	4	4	2	4	11	3
8	5	4	1		15	4
9	6	5	1		18	3
10	7	6	1		22	4
11	8	7	2	8	25	3
12	9	8	1		29	4
13	10	8	1		32	3
14	11	9	2	11	36	4
15	12	10	1		39	3
16	13	11	1		43	4
17	14	11	1		46	3
18	15	12	2	15	50	4
19	16	13	1		53	3
20	17	14	1		57	4
21	18	15	2	18	60	3
22	19	15	1		64	4
23	20	16	1		67	3
24	21	17	1		71	4
25	22	18	2	22	74	3
26	23	18	1		78	4
27	24	19	1		81	3
28	25	20	2	25	85	4

Fig. 7.9. Exploring patters stemming from the integer exponents 6^n.

Chapter 8

Historical Connections and Geometric Modeling

8.1 Introduction

This chapter focuses on the use of a spreadsheet as a tool for mathematical concept development through numerical modeling of grade-appropriate problems situated in a geometric context and connected to mathematics of antiquity. It underscores the importance of the historical component in the preparation of teachers by using a context that can be traced back to Babylonian mathematics (2000-1600 B.C.). Over this background, the chapter offers pedagogical ideas demonstrating how one can use algebraic inequalities and associated proof techniques in context-oriented, problem-solving situations. It has been argued [Conference Board of the Mathematical Sciences, 2001] and evidence has been provided [Abramovich, 2005, 2006, 2011] that the appropriate use of spreadsheets can motivate and encourage making connections between algebraic inequalities and mathematical modeling.

The focus of this chapter is consistent with standards for teaching mathematics [National Council of Teachers of Mathematics, 2000; Common Core State Standards, 2010; Ontario Ministry of Education, 2005a, 2005b; Western and Northern Canadian Protocol, 2008; Ministry of Education, Singapore, 2006, 2012] and recommendations for the preparation of the teachers of mathematics [Conference Board of the Mathematical Sciences, 2001, 2012; Advisory Committee on Mathematics Education, 2011]. In particular, some recommendations include the need for teacher candidates to understand "the ways that basic ideas of number theory and algebraic structures underlie rules for operations on expressions, equations, and inequalities" [Conference Board of the Mathematical Sciences, 2001, p. 40] as well as the importance of courses within which one "could examine the crucial role of algebra in use of computer tools like spreadsheets and the ways that computer algebra systems might be useful in exploring algebraic ideas" [*ibid*, p. 41]. Teacher candidates need to know how to use technology for it "contributes to a learning environment in which the curiosity of students can lead to rich mathematical discoveries" [Western and Northern Canadian Protocol, 2008, p. 9].

Toward this end, this chapter, in the context of spreadsheets, demonstrates how "the proper use of technology can make complex ideas tractable [and] can also help one understand subtle mathematical concepts" [Conference Board of the Mathematical Sciences, 2012, p. 57]. Also, the National Council of Teachers of Mathematics [2000] recommended that in grades 9-12 all students should "understand the meaning of equivalent forms of expressions, equations, inequalities and relations; write equivalent forms of equations, inequalities, and systems of equations and solve them with fluency … using technology in all cases" (p. 296) and it advocated for "instructional programs … [enabling] all students to select and use various types of reasoning and methods of proof" (p. 342). Teacher candidates "must know that proof and deduction are used not only to convince but also to solve problems and gain insights" [Conference Board of the Mathematical Sciences, 2012, p. 59]. In the classroom practice, such programs encourage teachers "to have the big picture in mind so that they can better understand … what they have to do at their level, as well as to plan and advise students in their learning of mathematics" [Ministry of Education, Singapore, 2012, p. 11]. These ambitious expectations for teaching mathematics and the curriculum of the subject matter, raising the level of professional standards for teachers from their current position, point at the importance of their training in conjecturing, proving, and using inequalities in the digital era.

Another reason to focus on inequalities in the context of teacher education stems from the importance of these tools in undergraduate mathematics curriculum. Teachers with little or no experience in using inequalities are not likely to adequately support their own students' long-term development. At the same time, teachers with this and similar experience are likely "to enable students who are interested in pursuing tertiary studies in mathematics, sciences and engineering to … develop thinking, reasoning, communication and modelling skills through a mathematical approach to problem solving and the use of mathematics language" [Ministry of Education, Singapore, 2012, p. 8]. This, in particular, should include the development of skills for the study of calculus where the need for the mastery of the "epsilon-delta" language and finding approximations to infinite structures in terms of inequalities is rife. In other words, teachers need "knowledge of the mathematics that students are likely to encounter when they leave high school for collegiate study" [Conference Board of the Mathematical Sciences, 2001, p. 122]. However, in their own undergraduate mathematics courses, teacher candidates are frequently missing a context in which problems involving inequalities arise. This is unfortunate because context, in general, plays an important role in one's conceptual development [Frykholm *et al.*, 2005; Murray, 2004; Núñez *et al.*, 1999].

Furthermore, even if teacher candidates understand how inequalities work in calculus, they may have difficulty in connecting this knowledge to pre-college mathematics content. Such a "vertical disconnect" [Cuoco, 2001] can be minimized through redesigning mathematics courses for prospective teachers to help them see connections between undergraduate mathematics they learn in college and school mathematics they will be teaching [Conference Board of the Mathematical Sciences, 2001, 2012]. Likewise, a note about the importance of developing teaching skills in making mathematical connections across the curriculum can be found in the Mathematical Needs Project [Advisory Committee on Mathematics Education, 2011] stressing that students do "need to be taught by a teacher who has a sound understanding of the connections in mathematics" (p. 2).

With this in mind, the potential of spreadsheet modeling as a context for teacher candidates' use of inequalities in computing applications can be utilized. First, a spreadsheet can be used as a generator of applied problems dealing with the computational efficiency of numerical modeling involved and leading to the use of inequalities. Such use of inequalities will include their formal demonstration through utilizing several proof techniques. Second, a spreadsheet can be used as an environment conducive to conjecturing of a classic inequality (though often unknown to teachers) that, in turn, can be used as a problem-solving tool outside the computational medium. Finally, a spreadsheet can motivate making connections between different concepts and in that way provide a learning environment within which the Connections standard and Technology principle [National Council of Teachers of Mathematics, 2000] can be meaningfully illustrated and presented to teachers as pillars of contemporary pedagogy of school mathematics.

Connecting mathematics to technology in the context of spreadsheets helps teachers understand what it means to "incorporate the mathematical capabilities, methods and questions that arise from use of all available technologies, especially those used in the workplace and those that are designed on mathematical principles" [Advisory Committee on Mathematics Education, 2011, p. 14].

8.2   Modeling as Problem Solving in a Historical Context

A notable change in curriculum and pedagogy of school mathematics over the last five decades is the focus on application-oriented modeling activities as a vehicle for teaching and learning mathematical concepts [Beberman, 1964; Blum, 2002; Lesh *et al.*, 2010]. Just as a problem-solving approach to mathematics teaching facilitates one's conceptual understanding through extracting these concepts from an appropriate context [Alfors, 1962; Freudenthal, 1978], mathematical modeling is conducive to learners' engagement in formulating and resolving problematic situations through the use of different models that describe these situations. Mathematical modeling provides one with "opportunities to apply mathematical problem-solving and reasoning skills to tackle a variety of problems … [something that further enables learning] to deal with ambiguity, make connections, select and apply appropriate mathematical concepts, identify assumptions and reflect on the solutions to real-world problems" [Ministry of Education, Singapore, 2012, pp. 15-16]. One can see, as it has been argued [Doerr and English, 2003; Nunokawa, 1995], that modeling and problem solving are closely related pedagogical strategies. The use of technology has great potential to enhance modeling pedagogy by having learners explore computer-based models, pose, and then solve problems that emerge from the need to improve these models.

From a didactic perspective, training in modeling pedagogy is ultimately structured by one's engagement in formulating and resolving problematic situations through the use of a variety of models that represent those situations [Saaty and Alexander, 1981]. This suggests a fundamental relationship that exists between modeling and problem posing. Furthermore, viewing problem solving and posing as two sides of the same coin [Brown and Walter, 1990; Cai *et al.*, 2015] suggests the importance of providing teachers with experiences in modeling through formulating, exploring, and resolving problematic situations that result in the discovery of new mathematical ideas and concepts.

While learning to use technology as a modeling tool, one can come across many computationally driven problematic situations the resolution of which requires the use of new concepts. As far as a spreadsheet is concerned, its computational nature enables immediate feedback so that one can test emerging strategies and see results in ways that were never possible with more traditional, paper-and-pencil materials. When such use of a spreadsheet is a part of technology-motivated mathematics teacher education coursework, the course instructor's role is to encourage prospective teachers to take intellectual risk through the formulation of mathematically meaningful questions about numerical patterns observed.

8.3 Spreadsheet as an Agent of a Mathematical Activity Leading to Inequalities

The use of geometry as a background and motivation for the study and development of algebraic ideas has a very long history. Around 1700 B.C., the Babylonians used geometric terms as a substitute for unknowns in algebraic problems [Kline, 1972]. For example, Babylonian algebra included systems of simultaneous equations the modern form of which is $xy + x - y = a$, $xy = b$ where x and y had been referred to as width and length (see Problem 8.2 below). This approach, in the absence of abstract symbols, made mathematical (algebraic) calculations more concrete. According to Burkert [1972], "the application of areas [by Babylonians] ... is primarily algebraic; they provide an equivalent for quadratic equations" (p. 454). In terms of contemporary didactics, geometry served as a context for introducing algebraic concepts through a numerical approach. Nowadays, this approach can be enhanced by technology. In the following sections of this chapter, it will be demonstrated how learning to move from novice practice to expert practice in spreadsheet modeling of simple geometric problems leads to the use of inequalities. Consider

Problem 8.1. *Find all rectangles with integer sides and perimeter 64 (linear units). Which rectangle has the largest area? Which rectangle has the smallest area?*

This (relatively simple) problem can be found among specific expectations for high school students in Canada taking a course on measurement and geometry [Ontario Ministry of Education, 2005b, p. 44]. It can be solved numerically within a spreadsheet by using different modeling techniques. For the purpose of introducing teacher candidates to the use of inequalities, one approach is particularly instructive. Given the value of perimeter, the software can identify the side lengths of the corresponding rectangles within a two-dimensional chart resembling the multiplication table. By displaying the products of these side lengths in the chart, the largest and the smallest such products (areas) can be found. The power of a spreadsheet is that it enables one to parameterize a problem by making any constant value involved a parameter. Keeping in mind that perimeter of any rectangle with integer sides is an even number, one can set x and y to represent integer sides (width and length) of a rectangle with perimeter $2p$ and use the equation

$$x + y = p \qquad (8.1)$$

as a mathematical model of Problem 8.1.

In the process of modeling Eq. (8.1) within a table, one faces the problem of finding ranges for variables x and y that determine the size of the table. Apparently, the larger the value of p, the larger is this size. For example, the spreadsheet pictured in Fig. 8.1 shows that even using a 400-cell (20×20) table does not allow one to find all rectangles of perimeter 64 as the table is limited to the side lengths not greater than 20 and the first missing rectangle has area 231 and side lengths 11 and 21 as the factorization $231 = 11 \cdot 21$ is the only one satisfying the condition $p = 32$. In such a way, a spreadsheet becomes an agent of the following mathematical activity (problem):

Given semi-perimeter p of a rectangle, determine the ranges within which its width x and length y vary.

Because this activity requires finding the smallest and the largest values for x and y, the result sought can be expressed in terms of inequalities.

The most basic inequality representing the order relationship between the width and the length can be technology-motivated. Indeed, numbers located diagonally within the spreadsheet of Fig. 8.1 repeat each other as x and y change their roles. Such a symmetric repetition can be eliminated if one assumes $x \leq y$ — the basic inequality relationship between the two variables. The next section shows how one can gain experience in developing more complicated inequalities using the inequality $x \leq y$ as the basic assumption about length and width.

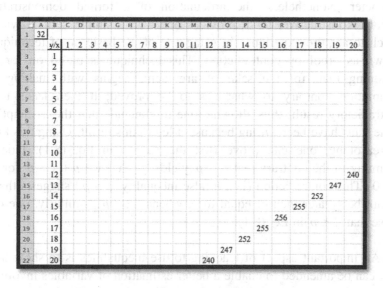

Fig. 8.1. Novice practice in spreadsheet modeling.

8.4 Developing Skills in Conjecturing, Proving, and Improving Inequalities

Following is a number of inequality-oriented propositions that can be conjectured and then proved by teacher candidates. These propositions will not only inform practice in spreadsheet modeling but their application will manifest a move from novice to expert practice. Besides the need to possess technological competence, teachers "need to see how reasoning and proof occur in high school mathematics outside of their traditional home in axiomatic Euclidean geometry" [Conference Board of the Mathematical Sciences, 2012, p. 59]. Also, they "need to know how to prove what is unstated in high school in order ... to be able to answer questions from students seeking further understanding" [*ibid*, p. 60]. Therefore, a spreadsheet environment allows teachers "to make use of original historical sources, which can motivate the theory and make it seem less disconnected from the school mathematics" [*ibid*, p. 60].

Proposition 8.1. In Eq. (8.1) both $x < p$ and $y < p$.

Proof. Note that the inequalities to be proved are quite obvious, especially when one is encouraged to interpret them in contextual terms: both width and length of a rectangle may only be smaller than its semi-perimeter. Nonetheless, the articulation of a formal demonstration (proof) is instructive, for it provides one with a simple context within which to acquire experience in using one of the basic proof techniques known as proof by contradiction. This technique is based on formal reasoning that, for the sake of argument, begins with making an assumption contrary to what has to be proved; it then leads to a contradictory result, thus allowing one to conclude that the assumption made must have been wrong because it led to this result. Using this kind of reasoning, one can prove first that $x < p$ in Eq. (8.1). Indeed, assuming to the contrary that $x \geq p$ yields $p = x + y \geq p + y > p$ for all $y \geq 0$. This contradiction (i.e., a false inequality $p > p$) suggests that x cannot be greater than or equal to p, whence $x < p$. Similarly, one can prove that $y < p$ in Eq. (8.1).

An important aspect of teaching to use inequalities is to show how they can be amended to enable a better estimation of variables involved; that is, "to determine a way to get from what is known to what is sought" [Western and Northern Canadian Protocol, 2008, p. 8]. Toward this end, teacher candidates can be encouraged to inquire: Are there less obvious (and, consequently, more accurate) estimates for x and y in Eq. (8.1) other than bounding them from above by p? Such an improvement in accuracy can be given by

Proposition 8.2. In Eq. (8.1), only one of the variables x and y can be greater than $p / 2$.

Proof. Once again, using proof by contradiction, one can make an assumption that $y \geq x > p / 2$ in Eq. (8.1). Therefore, $p = x + y > p / 2 + p / 2 = p$; that is, $p > p$. This contradiction suggests that if $x \leq y$ then $x \leq p / 2$. Similarly, one can prove that $y \geq p / 2$ in Eq. (8.1). Indeed, assuming $y < p / 2$ yields $p = x + y < p / 2 + p / 2 = p$, a contradiction.

Remark 8.1. The inequality $x \leq p/2$ is an improvement over that of $x < p$ as the set of points defined by the latter inequality includes the set of points defined by the former inequality but not vice versa. That is, the inequality $x \leq p/2$ does imply the inequality $x < p$; yet the inequality $x < p$ does not necessarily imply $x \leq p/2$. A similar interplay exists between the inequalities $y \geq p/2$ and $y < p$. Indeed, the inequality $y < p$ does not imply the inequality $y \geq p/2$ and, therefore, the latter inequality does improve the former one not allowing y to be smaller than $p/2$.

In turn, the inequality $x \leq p/2$ can be amended by recognizing that whereas p may be not a multiple of two, x is an integer variable. This leads to

Proposition 8.3. If $x \leq y$ in Eq. (8.1), then $x \leq INT(p/2)$.

Proof. By definition, $INT(z)$ is the largest integer smaller than or equal to z. In other words, $0 \leq z - INT(z) < 1$ for any real number z. Once again, in order to prove the inequality $x \leq INT(p/2)$, one can use proof by contradiction. To this end, one may assume that $x > INT(p/2) \neq p/2$. In this case, because x is an integer, the assumption $x > INT(p/2)$ implies that x may be at least equal to $INT(p/2)+1$; that is, $x \geq INT(p/2)+1 > p/2$ which contradicts the inequality $x \leq p/2$. For example, when $p = 5$ the condition for an integer x to be smaller than or equal to $5/2$ implies $x \leq INT(5/2) = 2$. Assuming to the contrary that $x > 2$, implies $x \geq 3$ and, in particular, $x > 5/2$. However, $x < 5/2$; otherwise, to the contrary of the equality $x + y = 5$, we would have $x + y \geq 2x > 5$.

Finally, the ranges for x and y in Eq. (8.1) can be established through

Proposition 8.4. If $x \leq y$ in Eq. (8.1), then

$$INT(p/2) \leq y \leq p-1, \quad 1 \leq x \leq INT(p/2) \qquad (8.2)$$

Proof. Let $x \leq y < INT(p/2)$. Then

$$p = x + y < 2 \cdot INT(p/2) \leq 2(p/2) = p.$$

This contradiction (i.e., a false inequality $p < p$) suggests that y cannot be smaller than $INT(p/2)$, thus $y \geq INT(p/2)$. Furthermore, the inequality $y \leq p - 1$ holds true. Once again, one can demonstrate the correctness of the last inequality using proof by contradiction.

Remark 8.2. Note that $INT(p/2) \leq p/2$; thus, the inequality $y \geq INT(p/2)$ is weaker than the inequality $y \geq p/2$.

Inequalities (8.2), being the most accurate ones among all that have been conjectured and proved above, can be used to construct a spreadsheet for numerical modeling of Eq. (8.1). Figure 8.2 shows such a spreadsheet in which the ranges for variables x and y depend on the slider-controlled value of semi-perimeter p set in cell A1. In that way, a spreadsheet becomes a *consumer* of mathematical activities carried out by teacher candidates through which certain algebraic inequalities have been conjectured and proved. One can see that for each pair (x, y) of side lengths satisfying inequalities (8.2), the spreadsheet displays the area of the corresponding rectangle. Furthermore, the described mathematical activities result in a computationally efficient spreadsheet. Indeed, unlike the case of a 400-cell table (Fig. 8.1), all solutions to Problem 8.1 have been displayed within even a smaller (16×16) table (Fig. 8.2). Analyzing these solutions, one can conclude that the square ($x = y = 16$) has the largest area and the rectangle with the smallest integer side length ($x = 1$) has the smallest area.

Using the results of Proposition 8.4, the spreadsheet of Fig. 8.3 is programmed as follows. In cell C2 the number 1 is entered; in cell D2 the formula =IF(C2=" "," ",IF(C2<INT($A$1/2),1+C2," ")) is defined and replicated across row 2. In cell B3 the formula =INT(A1/2) is defined; in cell B4 the formula =IF(B3=" "," ",IF(B3<$A$1-1,1+B3," ")) is defined and replicated down column B. In cell C3 the formula

=IF(OR(C$2=" ",$B3=" "), " ", IF(C$2+$B3=$A$1,C2*$B3," "))

is defined and replicated across rows and down columns. As a result, in the case of semi-perimeter $p = 32$, assuming that $x \leq y$ the spreadsheet sets the ranges [1, 16] and [16, 31] for the side lengths x and y, respectively, and generates 16 possible areas with 31 being the smallest ($x = 1, y = 31$) and 256 being the largest ($x = y = 16$).

	A	B	C	D	E	F	G	H	I	J	K	L	M	N	O	P	Q	R
1	32																	
2		y/x	1	2	3	4	5	6	7	8	9	10	11	12	13	14	15	16
3		16																256
4		17															255	
5		18														252		
6		19													247			
7		20												240				
8		21											231					
9		22										220						
10		23									207							
11		24								192								
12		25							175									
13		26						156										
14		27					135											
15		28				112												
16		29			87													
17		30		60														
18		31	31															

Fig. 8.2. Expert practice in spreadsheet modeling.

At this point, teacher candidates should be reminded that, in general, given a perimeter, no rectangle with the smallest area exists. An interesting exercise involving the use of inequalities is to prove that for any rectangle with semi-perimeter p and the smallest side x there exists a positive number $\varepsilon < x$ such that the relationship between the areas of the rectangles with the side lengths $(x, \frac{p}{2} - x)$ and $(x - \varepsilon, \frac{p}{2} - x + \varepsilon)$ is expressed by the inequality $(x - \varepsilon)(\frac{p}{2} - x + \varepsilon) < x(\frac{p}{2} - x)$. In other words, by allowing the rectangle to have non-integer sides, one can make its area smaller than any positive number, no matter how small the number is. The last inequality can be presented in the form

$$x\frac{p}{2} - x^2 + x\varepsilon - \varepsilon\frac{p}{2} + x\varepsilon - \varepsilon^2 < x\frac{p}{2} - x^2$$

the simplification of which yields $2x\varepsilon - \frac{p}{2}\varepsilon - \varepsilon^2 < 0$ whence $2\varepsilon(\frac{p}{4} - x + \frac{\varepsilon}{2}) > 0$ — a true inequality because $x \leq \frac{p}{4}$ and $\varepsilon > 0$.

Finally, the above computational discovery that a square provides the largest area for all the rectangles sharing the same perimeter is of a special importance, both from theoretical and applied perspectives.

Although not yet formally demonstrated, this discovery can serve as a motivation for further learning to use inequalities; in particular, to facilitate such a demonstration. As noted by Kline [1985], "A farmer who seeks the rectangle of maximum area with given perimeter might, after finding the answer to his question, turn to gardening, but a mathematician who obtains such a neat result would not stop there" (p. 133). The next section shows how this characteristic feature of the mathematical mind can be put to work in the context of spreadsheet modeling and use of inequalities.

8.5   Technology-Motivated Conjecturing

As it was previously mentioned, using a spreadsheet as a tool for conjecturing is an example of using technology as a cultural amplifier of mathematical activities. Another example of such cultural amplification is the use of a hand-held calculator that enhances human ability to carry out arithmetical computations. However, unlike the calculator, the spreadsheet of Fig. 8.2 is a goal-oriented tool developed by someone who, by using inequalities, progressed from novice to expert practice in spreadsheet modeling. This distinction represents an important feature of the spreadsheet-enabled mathematics pedagogy, manifesting a qualitatively different perspective on the use of technology in comparison with approaches focusing on calculations alone. Here a learner not only demonstrates skills in using a spreadsheet as an amplifier, but better still, uses higher-order skills to amplify the amplifier. In the words of Newman *et al.* [1989], a "second-level amplification" (p. 130) takes place through the use of inequalities in spreadsheet modeling.

Indeed, the modeling data generated by the spreadsheet can be utilized to conjecture new relationships between variables involved. With this in mind, consider the arrangement of numbers (forming a right triangle) displayed in the spreadsheet of Fig. 8.2. As one moves up along the hypotenuse filled with the values of areas in the range [31, 256], the excess of length y over width x becomes smaller and smaller. In other words, the smaller the difference between two adjacent sides of the rectangle, the bigger its area is. Furthermore, for each number located on the hypotenuse, the sum of the corresponding numbers on the legs is invariant across all the triples and is equal to the given semi-perimeter.

As an example, consider the triple (175, 25, 7), the elements of which are located in cells I12, B12, and I2, respectively (Fig. 8.2). The sum of the second and third elements of the triple is exactly the given semi-perimeter of the rectangle. Also, for the triple (192, 24, 8) we have 24 − 8 < 25 − 7. The same two phenomena can be observed for other values of the semi-perimeter when changing the value of cell A1.

In addition, proceeding from numerical evidence generated by the spreadsheet of Fig. 8.2, one can come up with the following conjecture: the square of the half of the sum of the width and length of the rectangle is greater than or equal to the area of this rectangle. In algebraic terms, this conjecture can be written in the form of the inequality

$$\left(\frac{x+y}{2}\right)^2 \geq xy \tag{8.3}$$

which turns into the equality when rectangle turns into square ($x = y$). An equivalent form of (8.3) is the inequality

$$\frac{x+y}{2} \geq \sqrt{xy}, \tag{8.4}$$

one of the most famous in mathematics known as the arithmetic mean-geometric mean (AM-GM) inequality. It holds true for all non-negative x and y, turns into equality when $x = y$, and can be generalized to any number of variables [Korovkin, 1961]. The inequality was mentioned by the Conference Board of the Mathematical Sciences [2012] among "advanced topics [applicable] to optimization problems" (p. 67). However, the author's experience indicates that many prospective teachers are not familiar with this very useful inequality, its various interpretations, applications, and generalizations.

In pre-calculus, inequality (8.4) is often used to find the largest or the smallest values of various characteristics of geometric figures thus avoiding the need for differentiation. For example, the following two statements (the first of which was already discovered computationally) can be immediately verified through the direct application of inequality (8.4): (a) among all rectangles of perimeter $2p$, square has the largest area equal to $p^2 / 4$ and (b) among all rectangles of area A, the smallest semi-perimeter is $2\sqrt{A}$. Indeed, in order to prove

(a) and (b) one has to write, respectively, $A = xy \leq (\frac{x+y}{2})^2 = \frac{p^2}{4}$ and

$p = x + y \geq 2\sqrt{xy} = 2\sqrt{A}$. In order to prove inequality (8.4), one has to present the difference between its left- and right-hand sides as a non-negative expression $\frac{1}{2}(\sqrt{x} - \sqrt{y})^2$ which is equal to zero when $x = y$.

8.6 Spreadsheet Modeling vs. Using AM-GM Inequality

Teachers' experience in using spreadsheets may include the discussion of the interplay that exists between computational and theoretical approaches to problem solving. In this section, using a problem from the Babylonian mathematics, it will be shown how the AM-GM inequality can be a better exploratory tool than a computational experiment. Such a comparison intends to demonstrate how theory can complement and refine numerical evidence. This is consistent with the Common Core State Standards [2010] emphasis on the importance "of creating a coherent representation of the problem at hand ... attending to the meaning of quantities, not just how to compute them" (p. 6). To this end, consider a problem from the cuneiform text AO 8862 expressed by the Babylonians in the following rhetorical form [Van der Waerden, 1961, p. 63]:

Problem 8.2. *"Length, width. I have multiplied length and width, thus obtaining the area. Then I added to the area, the excess of the length over the width ...183 was the result ... Required length, width and area".*

Retaining the meaning of variables x and y introduced in Sec. 8.3, one can present the mathematical model of Problem 8.2 in the form of the equation $xy + y - x = 183$, or, making its right-hand side a parameter, more generally,

$$xy + y - x = n. \qquad (8.5)$$

In particular, (non-linear) Eq. (8.5) shows that the smaller the excess of the length over the width, the bigger the area of the rectangle is. This fact was already discovered experimentally by analyzing the data resulting from modeling (linear) Eq. (8.1). In this way, one can see how the results of a computational experiment with a linear mathematical model can be interpreted through a theoretical analysis of a non-linear mathematical model.

Equation (8.5) can be solved numerically within a two-column chart for any given value of n. Indeed, the inequality $x \le y$ in combination with Eq. (8.5) yields the inequality $x^2 \le n$. To this end, Eq. (8.5) can be rewritten in the equivalent form

$$y = (n + x) / (x + 1). \tag{8.6}$$

Consequently, the inequality $(n + x) / (x + 1) \ge x$ is equivalent to $x^2 \le n$. Alternatively, one can prove the following two propositions using different reasoning techniques — proof by contradiction and direct proof. The latter technique is a combination of earlier established (and often simpler) facts leading to a new mathematical result.

Proposition 8.5. In Eq. (8.5), the inequality $x \le y$ implies $x \le \sqrt{n}$.

Proof. Assuming, on the contrary, that $x > \sqrt{n}$ and using Eq. (8.6) yields $y = 1 + (n - 1) / (x + 1) < 1 + (n - 1) / (\sqrt{n} - 1) = \sqrt{n}$; that is, $y < \sqrt{n}$. This, however, contradicts the inequality $x \le y$.

Proposition 8.6. In Eq. (8.5), the inequality $x \ge 1$ implies $y \le (n + 1) / 2$.

Proof. This time one can use direct proof and estimate y defined through established earlier equality (8.6). To this end, one can use a combination of an equivalent representation of the right-hand side of (8.6) and the inequality $x \ge 1$ as follows:

$$y = 1 + (n - 1) / (x + 1) \le 1 + (n - 1) / 2 = (n + 1) / 2.$$

Figure 8.3 (left) shows a spreadsheet that computes values of y according to equality (8.6) for $n = 183$ with x varying from 1 to $\sqrt{n}$. It turns out that for $n = 183$, there are two rectangles, (12, 15) and (13, 14), with perimeter 54, the smallest perimeter among all possible rectangles with integer sides that satisfy Eq. (8.5). Changing n to 182 in Fig. 8.3 (right) shows no solution to Eq. (8.5), meaning that for $n = 182$, the graph of the corresponding equation does not contain points with integer coordinates for $x \le y$.

Using a circular reference in a spreadsheet formula enables one to construct a table (see the range H2:I12 in the spreadsheet of Fig. 8.4) that shows how many equations within the n-range [1, 200] have a specified number of integer solutions.

	A	B
1	$n=$	183
2		
3	x	y
4	1	92
5	2	
6	3	
7	4	
8	5	
9	6	27
10	7	
11	8	
12	9	
13	10	
14	11	
15	12	15
16	13	14
17	14	

	A	B
1	$n=$	182
2		
3	x	y
4	1	
5	2	
6	3	
7	4	
8	5	
9	6	
10	7	
11	8	
12	9	
13	10	
14	11	
15	12	
16	13	
17	14	

Fig. 8.3. Modeling Eq. (8.5) within a two-column chart.

Below are the programming details related to Fig. 8.4. Cell A4: =1; cell A5: =IF(A4=" "," ",IF(A4^2<B$1,A4+1," ")) — replicated down column A; cell B4:
=IF(A4=" "," ", IF(AND(INT((B$1+A4)/(A4+1)) = (B$1+A4)/(A4+1), (B$1+A4)/(A4+1)>=A4), (B$1+A4)/(A4+1)," ")) — replicated down column B; cell C1, controlled by a slider, includes a ratio (mentioned in the next formula) of the side lengths of a rectangle; cell C4: =IF(OR(B5=" ",A5=" "), " ", IF(A5=B5,"SQUARE", IF(B5=C$1*A5, "RATIO"," "))) — replicated down column C; cell D4: =IF(B4=" "," ",B4-A4) — replicated down column D; cell E1: =COUNT(B4:B20); cell E4: =IF(B4=" "," ", A4*B4) — replicated down column E; column F is filled with consecutive natural numbers starting from one; cell G1: =IF(A$1=0, " ", IF (A$1=F1,E$1, G1)) — this formula includes a circular reference and is replicated down column G; entries in column H represent a possible number of solutions to Eq. (8.5); in cell I2 the formula =COUNTIF(G$1:G$200,H2) is defined and replicated down column I. As one varies n (i.e., the content of cell B1) from 1 to 200 by playing the slider attached to cell B1, the spreadsheet generates the number of solutions to Eq. (8.5) for each value of n and interactively calculates in

column I the sums of each number of solutions. The dependence on the number of solutions to Eq. (8.5) of the total number of equations in the n-range [1, 200] with that number of solutions is shown in the chart embedded into the spreadsheet of Fig. 8.4.

In particular, one can discover that more than 50% of n-values in this range provide at most one integer solution to Eq. (8.5). Also, no value of n provides more than eight solutions to Eq. (8.5); that is, the maximum number of rectangles satisfying Problem 8.2 within the n-range [1, 200] is eight.

	A	B	C	D	E	F	G	H	I
1	n=	200	2	xy+y-x=a	0	1	1	number of solutions	number of equations
2		•				2	0	0	45
3	x	y	type	excess	area	3	1	1	63
4	1					4	1	2	30
5	2					5	1	3	29
6	3					6	0	4	13
7	4					7	2	5	11
8	5					8	0	6	5
9	6					9	2	7	1
10	7					10	1	8	3
11	8					11	1	9	0
12	9					12	0	10	0
13	10								
14	11								
15	12								
16	13								
17	14								
18	15								
19									
20									

Fig. 8.4. Exploring the number of solutions to Eq. (8.5) in the range $1 \le n \le 200$.

Remark 8.3. One can use *Wolfram Alpha* to find integer solutions of Eq. (8.5). To this end, one can enter the following request into the input box of the program: "solve in integers $xy + y - x = 183$, $y >= x > 0$". As a result, the program yields the same four solutions shown in Fig. 8.3 (left). One may wonder as to why we need a spreadsheet with its relatively complicated programming when *Wolfram Alpha* can be used easily to find all the solutions? While this kind of curiosity makes sense, one has to keep in mind that the unity of spreadsheets, historical perspectives, and plane geometry was proposed as a learning environment for using algebraic inequalities as tools in computing applications. At the same time, integrating a spreadsheet with *Wolfram Alpha* enables one to verify the (formally proved) results of such use of inequalities by accurately formulating a problem for the software to solve. Furthermore, teachers and their students alike can use *Wolfram Alpha*, just as professional mathematicians can use any proof assistant technology, as a means "to put the correctness of their proofs beyond reasonable doubt" [Harrison, 2008, p. 1405].

Remark 8.4. One of the recommendations for teachers in Australia [National Curriculum Board, 2008] suggests "to extend students in more depth in key topics … [so that] advanced students … could be posed a question like: 'Can you describe some shapes that have the same number of perimeter units as area units?' … [thus being offered] opportunities for examination of a range of shapes, for use of algebraic methods, and even the historical dimension of this problem" (p. 8). To this end, one can use one of the spreadsheets (either Fig. 8.3 or 8.4) and find those values of n for which a rectangle possesses the property described in the above-cited question. In doing so, one can find that there are only two values of n, namely $n = 16$ and $n = 21$, for which such rectangles exist. As mentioned in the following quote from Plutarch[o] [Van der Waerden 1961, p. 96]:

The Pythagoreans also have a horror for the number 17. For 17 lies exactly halfway between 16, which is a square, and the number 18, which is the double of a square, these two being the only numbers representing areas, for which the perimeter (of the rectangle) equals the area.

More specifically, $n = 16$ relates to the first figure (a four by four square) and $n = 21$ — to the second figure (the double of a three by three square or a three by six rectangle). Alternatively, one can construct a spreadsheet (or use *Wolfram Alpha*) to solve the equation $xy = 2(x + y)$ in positive integers to get the following two pairs: $(x, y) = (4, 4)$ and $(x, y) = (3, 6)$.

---

[o] Plutarch — a Greek historian of the first-second centuries A.D.

Remark 8.5. One may be reminded that the generalized Problem 8.2 defined by Eq. (8.5) calls for integer values of width and length and, therefore, the absence of solutions for some values of n, as well as the presence of two rectangles with the same perimeter for other values of n, should be interpreted correctly. Indeed, these findings are true only when Eq. (8.5) is considered a Diophantine equation — an indeterminate equation with integer coefficients that has to be solved in integers. When this constraint is removed, Eq. (8.5) reveals a different structure of its solutions. With this in mind, extending the generalized Problem 8.2 to allow for non-integer values of width and length, one can find that for each value of n, one and only one rectangle has the smallest perimeter. This fact can be established theoretically by using the AM-GM inequality. To this end, one can write

$$y + x = \frac{n+x}{x+1} + x + 1 - 1 = x + 1 + \frac{n-1}{x+1} \geq 2\sqrt{(x+1)\frac{n-1}{x+1}} = 2\sqrt{n-1}.$$

Therefore, the value of the smallest perimeter is $4\sqrt{n-1}$. Furthermore, the equality (i.e., the case $y + x = 2\sqrt{n-1}$) takes place when $x + 1 = \frac{n-1}{x+1}$ yielding $x = \sqrt{n-1} - 1$ and $y = \sqrt{n-1} + 1$, the width and length of the rectangle with the smallest perimeter satisfying Eq. (8.5) extended to any values of x and y. The area xy of such a rectangle is equal to $(\sqrt{n-1} - 1)(\sqrt{n-1} + 1) = n - 2$ and the length y is two more than the width x. One may wonder as to why, whereas some values of n do not provide rectangles with integer sides, the area is always an integer. Why do rectangles with integer area not satisfy conditions of the generalized Problem 8.2? Consider, for example, $n = 180$. There is a rectangle with area 178 that has integer width $x = 2$ and length $y = 89$, yet these values of x and y differ by more than two and do not satisfy Eq. (8.5) for that value of n.

Remark 8.6. Coming back to the integer domain, one may be asked to define those values of n for which $n - 2$ can be factored into two integers with difference two. In doing so, one can recognize that $n - 1$ should be a perfect square (to allow for integer values of $\sqrt{n-1} \pm 1$) and, therefore, n should be one more than a square number. This fact could first be discovered through numerical evidence by using the spreadsheet of Fig. 8.3. For example, the spreadsheet would show that $n = 197$ (= $14^2 + 1$) yields a rectangle, the integer sides of which differ by two, and this rectangle is unique. It has the smallest perimeter out of all rectangles satisfying Eq. (8.5). Exploring such a question from a theoretical perspective develops appropriate mathematical experience in aiding reflective inquiry and making mathematical connections.

To conclude this section, note that technology-motivated conjecturing of the AM-GM inequality and its subsequent use enabled one to solve the first part of Problem 8.1 (seeking the largest area) not only without using technology but better still, the approach avoids a possible misinterpretation of the modeling data. Furthermore, it allows one to participate in a theoretical reflection on modeling data, which is a way of giving legitimacy to the model used. It is in this sense a spreadsheet can serve as a cultural amplifier of mathematical activities enabling the discovery of theoretical knowledge that complements pure computational problem-solving techniques. This amplification, however, is not straightforward in a sense that it requires certain reorganization of mathematical activities making possible the application of theory and the interpretation of experimental data.

In this way, theoretical knowledge allows one to obtain results that may be overlooked within a computational experiment. By the same token, a computational experiment may bring new knowledge requiring conceptualization. It is important for teacher candidates to appreciate the interplay between theoretical and experiential knowledge in mathematics. Such an appreciation helps them "to further develop the habits of mind that define mathematical approaches to problems" [Conference Board of the Mathematical Sciences, 2001, p. 46].

8.7 Making Connections as a Good Teaching Practice

The Conference Board of the Mathematical Sciences [2001] recommended the development of courses for prospective teachers that include opportunities to "look deeply at fundamental ideas, to connect topics that often seem unrelated" (p. 46). The courses "must take into cognizance the new generation of learners, the innovations in pedagogies as well as the affordances of technologies" [Ministry of Education, Singapore, 2012, p. 2]. Likewise, the courses should support "building and guiding mathematical reasoning — generalizing, finding common structures in theorems and proofs ... [in order to] make it easier to organize and understand mathematical ideas" [Conference Board of the Mathematical Sciences, 2012, p. 56]. In other words, the main role of such courses is to develop learning opportunities of using mathematical and technological knowledge in support of effective teaching of the subject matter [Ball, 2000].

The context of problems on perimeter and area explored by combining computational and theoretical approaches provides teacher candidates with ample opportunities to deepen their knowledge of mathematics within the technological pedagogical content knowledge framework [Niess, 2005; Mishra and Koehler, 2006] developed at the confluence of Shulman's [1986, 1987] notion of pedagogical content knowledge and that of Maddux's [1984] Type I/Type II technology applications. The appropriate use of this framework facilitates making connections between seemingly disconnected concepts, something that "helps students make sense of what they learn in mathematics" [Ministry of Education, Singapore, 2006, p. 4]. For example, one can be encouraged to make a connection between what was discovered through analyzing the spreadsheet of Fig. 8.3 and Eq. (8.5). Indeed, it was discovered that the smaller the difference between the length and the width of a rectangle, the larger the area of the rectangle. But that is exactly what Eq. (8.5) represents.

Teacher candidates may be encouraged to find out if the two models are identical. In particular, they may be asked what values of n in Eq. (8.5) provide a square. In much the same way, this inquiry can be extended to include rectangles with the length and width being in a given ratio. (This explains the meaning of the formula defined in column C (beginning from cell C4) presented above through the description of the programming of the spreadsheet of Fig. 8.4). By using a spreadsheet, one can discover that the values of n for which the length is twice the width represent a sequence of consecutive triangular numbers of even rank starting from three. One can then be asked to explain this phenomenon algebraically by substituting $y = 2x$ in Eq. (8.5). In doing so, one can discover that testing if the integer n is a triangular number of an even rank $2x$ [Abramovich *et al.*, 1995], leads to the Diophantine equation $x(2x+1) = n$ which is equivalent to Eq. (8.5) having an integer solution when $y = 2x$.

Furthermore, one can use a spreadsheet to discover the values of n for which the length can be three, four, five, *etc.* times as much as the width. Learning first hand this experimental approach to mathematics has the potential to develop one's confidence in taking intellectual risk to move beyond the bounds of the given curriculum in the digital era. In other words, the more teacher candidates have experience in recognizing connections among different concepts through participating in sophisticated computational experiments, the more they can help their students, using a wide range of mathematical representations and modeling techniques, to "pursue mathematical ideas in addition to solving the problem at hand" [National Council of Teachers of Mathematics, 2000, p. 359].

Chapter 9

Computational Problem Solving in Context

9.1 Introduction

A number sequence in which the difference between any two consecutive terms is a constant is called an arithmetic sequence. The simplest example of such a sequence is the natural number sequence 1, 2, 3, 4, ..., where the difference between any two consecutive terms is one. Similarly, one can consider the sequence of consecutive odd (1, 3, 5, 7, ...) or even (2, 4, 6, 8, ...) numbers as arithmetic sequences with the difference two.

An arithmetic sequence is one of the most basic entities of algebra and number theory. In Chapter 4, in the context of prime numbers, a spreadsheet was used to find five sequential primes (e.g., 5, 11, 17, 23, 29) in arithmetic progression with difference six, and five sequential primes (e.g., 347, 349, 353, 359, 367) with gaps (2, 4, 6, 8) in arithmetic progression with difference two. That is, the property of numbers being in arithmetic progression can be observed as the fundamental description of some elements of a complex mathematical phenomenon. The exploration of arithmetic sequences lends itself to the joint use of context and technology. The use of spreadsheets in teaching topics in algebra and number theory to students at pre-college level and to their future teachers of mathematics has been well documented in the literature [Abramovich, 1999; Abramovich and Brantlinger, 2004; Baker, 2007, 2013; Calder, 2010; Haspekian, 2005; Hoyles, 1994; Sugden, 2005; Sugden and Miller, 2010; Sutherland, 1991; Sutherland and Rojano, 1993; Tabach and Friedlander, 2008]. The current approach to teaching mathematics in context confronts mathematics educators with a task of representing traditional number theory concepts in informal way, something that gradually develops a path to mathematical ideas through their numerical representation in computational environments. In this chapter, such an approach is suggested in the form of number of "authentic" situations supported by the use of a spreadsheet.

To begin, note that many mathematical problems lead to the summation of consecutive natural, odd, or even numbers. For example, the need to find the sum $1+2+3+...+n$ of the first n natural numbers emerges in different contexts such as finding the sum of all numbers in the $n \times n$ addition table (Chapter 2), finding the sum of all numbers in the $n \times n$ multiplication table or the number of rectangles within the $n \times n$ checkerboard [National Council of Teachers of Mathematics, 2000], finding the number of cubes or right rectangular prisms within the $n \times n \times n$ Rubik's cube [Abramovich, 2011], counting the number of handshakes among $n+1$ people [Steele and Johanning, 2004], finding the number of ways frequencies can be assigned to $n+1$ transmitters in a problem of cellular telephone network [Abramovich and Lyandres, 2009].

In particular, the number of handshakes among $n+1$ people can be interpreted mathematically as the number of ways of dividing a set of $n+1$ people into n non-empty subsets. To clarify, note that each person can be given one of the labels from the set $\{1, 2, 3, ..., n, n+1\}$. Then the first n people can be uniquely put in n groups and then the $(n+1)$-st person can join them in n ways, thus making n handshakes. This person, after making all possible handshakes, may not be considered any more in the process of counting handshakes. Now, the first $n-1$ people can form uniquely $n-1$ groups and the n-th person can join them in $n-1$ ways, thus making $n-1$ new handshakes. Again, this person has made all possible handshakes and, thereby, may leave. This process continues until only two people remain, thus making the last handshake. In that way, the sum $1+2+3+...+n$, which lists the handshakes in the order opposite to the one described above, represents the total number of handshakes among $n+1$ people.

Finding the sums of that kind in context can be extended to the summation of other arithmetic sequences. Partial sums of arithmetic sequences (e.g.: 1, 1 + 2 = 3, 1 + 2 + 3 = 6, 1 + 2 + 3 + 4 = 10, ...; 1, 1 + 3 = 4, 1 + 3 + 5 = 9, 1 + 3 + 5 + 7 = 16, ...; 1, 1 + 4 = 5, 1 + 4 + 7 = 12, 1 + 4 + 7 + 10 = 22, ...) form new number sequences known as polygonal numbers. In particular, these numbers, having strong connection to geometry, include triangular (1, 3, 6, 10, ...) and square (1, 4, 9, 16, ...) numbers which can be found in the school curriculum as links between different mathematical concepts [e.g., Hoyles, 1994; New York State Education Department, 1996; Szetela, 1999, Knuth, 2002; Palatnik and Koichu, 2014].

In the last two decades, triangular and square numbers as concepts appropriate for K-12 curriculum appeared in a number of mathematics education publications around the world [Abramovich *et al.*, 1995; Abramovich and Sugden, 2008; Ben-Chaim *et al.*, 1989; Hitt, 1994; Hoyles, 1994; Pugalee, 2001; Szetela, 1999], including standards for teaching mathematics [Department for Education, 2014; National Council of Teachers of Mathematics, 2000; New York State Education Department, 1996]. This prompts the idea of embedding these concepts in a grade-appropriate context enhanced by a spreadsheet as a means of motivating mathematical learning. Such a unity of context, mathematics, and technology makes it possible for triangular, square, and, more general, polygonal (*n*-gonal) numbers to "be well motivated so that they are seen as helpful, powerful tools that make it easier to organize and understand mathematical ideas" [Conference Board of the Mathematical Sciences, 2012, p. 56].

9.2 Establishing Context for Triangular Numbers

In what follows, the use of a spreadsheet will be considered, using Pollak's [1970] terminology, in a whimsical context. While it may be argued that problems of whimsy have only superficial connection to the real world, the accepted use of the term modeling embraces all possible relations between mathematics and the world outside it [Blum, 2002]. It appears that one's perception of what is whimsical and what is not largely depends on one's experience; in fact, many of today's real-life situations seemed like fictions yesterday. Furthermore, the strong relationship that exists between modeling and problem solving suggests the importance for the spreadsheet modeling to be carried out in a whimsical context that very often provides a powerful cognitive milieu for solving problems.

For example, in one of the first educational papers on mathematical modeling and applications, Engel [1968] explored such contexts as fishing, book reading, and coin tossing. It has been shown that each of these 'whimsical' contexts is conducive for using modeling strategies as means of instruction and developing the habit of seeing possible applications of mathematics. Note that context itself does not account for the mathematical content — the latter usually begins with a quantitative inquiry into the former, something that may be referred to as mathematization. We begin with

Brain Teaser 9.1: *Once upon a time, Marriott built a hotel in San Diego shaped as a lengthened parallelepiped with more than 100 rooms on each story overlooking the waterline of the ocean. The 13th story turned out to be a dangerous place to stay: every night a ghost visited a room there. It was observed that the ghost started with room 1, on the second night he emerged in room 3, on the third night he emerged in room 6, then he emerged in room 10, and so on. After a few such nights, the hotel's manager hired a secondary mathematics teacher candidate to investigate the behavioral pattern of the ghost by using a spreadsheet. More specifically, the manager wanted to know:*

Which room would the ghost visit on the 7th, 8th, and 9th nights?
Would room 100 be a ghost-free room?
If room 100 would not be a ghost-free room, which room might the ghost visit on the first night in order to end up in room 100?

9.3   Using a Spreadsheet to Investigate Brain Teaser 9.1

By analyzing the brainteaser, one can notice that after the first night the ghost skips one room, after the second night — two rooms, after the third night — three rooms, and so on. In such a way, the night number and the number of rooms skipped by the ghost on that night appeared to be equal to each other. In a spreadsheet environment, a move from one natural number to another (or, alternatively, a move from room to room) can be described both in terms of action and operation. For example, the move from number one to number three (alternatively, from room 1 to room 3) can be described both as skipping one room (an action) and adding two to one (an operation). Similarly, move from room 3 to room 6 can be described as skipping two rooms (an action) and adding three to three (an operation). Both the action and the operation are equivalent descriptions of the ghost's move from room to room. In general, the action of skipping n consecutive rooms after visiting room number k can be formally described as the operation $k + (n + 1)$ which yields the next room number to be visited by the ghost.

Whereas the action-oriented description is informal in a sense that it cannot be computerized, the description in terms of operation bears formalization needed for the subsequent spreadsheet programming. Figure 9.1 shows the spreadsheet in which a formula that generates numbers in row 3 (room numbers visited by the ghost) is constructed by conceptualizing the moves of the ghost in terms of addition and using natural numbers (row 2) as mediators. To this end, the formula =A3+B2 can be defined in cell B3 and replicated across row 3. This formula generates the sequence of room numbers by adding a current night number to a room number associated with the ghost's visit on the previous night. In such a way, the spreadsheet pictured in Fig. 9.1 generates the sequence of triangular numbers

$$1, 3, 6, 10, 15, 21, 28, 36, 45, \ldots . \qquad (9.1)$$

	A	B	C	D	E	F	G	H	I	J	K	L	M	N	O
1	☠		☠			☠				☠					☠
2	1	2	3	4	5	6	7	8	9	10	11	12	13	14	15
3	1	3	6	10	15	21	28	36	45	55	66	78	91	105	120

Fig. 9.1. Triangular numbers (row 3) generated by a spreadsheet.

One can check to see that the second difference of sequence (9.1) is equal to one. Indeed, the sequence of the first differences is 2 (= 3 − 1), 3 (= 6 − 3), 4 (= 10 − 6), 5 (= 15 − 10), 6 (= 21 − 15), ... ; the sequence of the second differences is 1 (= 3 − 2), 1 (= 4 − 3), 1 (= 5 − 4), 1 (= 6 − 5), ... Geometrically, sequence (9.1) can be represented through the dot diagram of Fig. 9.2 in which triangles evolve from a single dot in the bottom-left corner so that each new base has one more dot than the previous base. This geometric interpretation explains the origin of the term triangular numbers.

The development of sequence (9.1) enables one, using empirical induction, to answer the first question posed by the manager: on the 7th, 8th and 9th nights the ghost shall visit, respectively, rooms 28, 36 and 45. In mathematical terms, numbers in row 2 (that is, night numbers) represent the ranks of the corresponding triangular numbers in row 3 (that is, room numbers) of the spreadsheet of Fig. 9.1. Put another way, night numbers are the ranks of the corresponding dangerous (non ghost-free) room numbers. A ghost with that type of behavior will be referred to below as triangular ghost (TG).

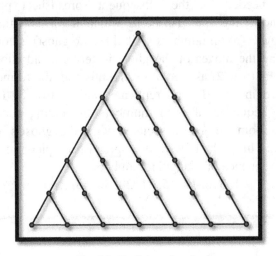

Fig. 9.2. Evolving triangles.

The second question posed by the manager about room 100 can also be answered just by looking at the spreadsheet of Fig. 9.1 — there is no such number in row 3. But one may be curious: What about rooms with numbers greater than 100 that are not displayed in Fig. 9.1? Such an investigation could be done by the spreadsheet pictured in Fig. 9.3. For example, in order to check electronically if room 100 is a ghost-free room, one has to enter the number 100 in cell A5, define the formula =COUNTIF(A3:GU3, A5) in cell B5 covering the range of 200 rooms, and, finally, define the formula =IF(B5=0, "Ghost Free", "Ghost") in cell C5. Then, the first room that the ghost could visit can become a slider-controlled variable (defined in cell A2) enabling one to find out whether there is a room from which the TG could begin visits in order to end up in room 100. In exploring the last inquiry one can come up with the following sequence of numbers representing rooms visited by the ghost on the first night with the goal to end up in room 100:

10, 23, 35, 46, 56, 65, 73, 80, 86, 91, 95, and 98.

More specifically, room 10 produces the sequence 10, 12, 15, ..., 100; room 23 produces the sequence 23, 25, 28, ..., 100; room 35 produces the sequence 35, 37, 40, ..., 100; and so on.

	A	B	C	D	E	F	G	H
1	☠		☠			☠		
2	1	2	3	4	5	6	7	8
3	1	3	6	10	15	21	28	36
4								
5	100	0	Ghost free					
6								

Fig. 9.3. Deciding the status of room 100.

In addition, one can attach a slider to cell A5 and carry out a similar investigation by changing the content of this cell. In other words, by using the spreadsheet shown in Fig. 9.3, one can check if any given room number is a ghost-free room or not. If the number 1 appears both in cells A3 and B5, a number that the spreadsheet generates in cell A5 is a triangular number; otherwise, this number is not a triangular number and the corresponding room is ghost-free (Fig. 9.3). Alternatively, in doing these kinds of exploration, one can use the square test for triangular numbers (see Sec. 9.6 below).

9.4 Developing Formulas for Triangular Numbers

The representation of triangular numbers in the (numeric) form of sequence (9.1) is insufficient to suggest any formal rule for generating these numbers through a mathematical definition (algebraic formula). However, numerical evidence provided by sequence (9.1) and the corresponding spreadsheet can be used to motivate the development of two distinct definitions for triangular numbers: recursive and closed. The fact that triangular numbers develop through a more complicated rule than an arithmetic sequence is reflected in the spreadsheet formula defined in cell B3 (Fig. 9.1). The formula contains relative references to two cells. The first reference, A3, relates to the previous room number (triangular number), the second reference, B2, relates to the current night number (a variable rate of change of dangerous room numbers). This suggests the following (recursive) rule for calculating the room number t_n on the night n: add current *night* number (n) to the *room* number visited on the previous night (t_{n-1}), keeping in mind that room 1 was visited on the first night. Symbolically, this rule can be expressed as

$$t_n = t_{n-1} + n, \ t_1 = 1. \tag{9.2}$$

Equation (9.2) represents the recursive definition of triangular numbers in the form of a difference equation of the first order. Note that by defining the seed value $t_1 = 1$, one starts using index n beginning from $n = 2$ which is consistent with the fact that the sequence of differences between consecutive triangular numbers begins with the number two. In other words, the room number corresponding to the first night is always one and the TG's decision about the next room number becomes clear on the second night.

In order to derive the closed formula for triangular numbers, one can note that any number in row 3 (Fig. 9.1) can be represented as the sum of its own rank and the ranks of all smaller numbers in this row. This observation can be confirmed by computing as follows: in the spreadsheet of Fig. 9.3 the formula =SUM($A2:B2) is defined in cell B3 and replicated across row 3. In that way, in cell H3 the last formula has the form =SUM($A2:H2), thereby suggesting the equality

$$36 = 8 + 7 + 6 + 5 + 4 + 3 + 2 + 1$$

which, in turn, prompts the following conjecture

$$t_n = 1 + 2 + 3 + \ldots + n. \tag{9.3}$$

B2				$f\!x$	=SUM($A1:B1)									
	A	B	C	D	E	F	G	H	I	J	K	L	M	N	O
1	1	2	3	4	5	6	7	8	9	10	11	12	13	14	15
2	1	3	6	10	15	21	28	36	45	55	66	78	91	105	120

Fig. 9.4. Discovering formula for the sum of the first n natural numbers.

Relation (9.3), however, is not a closed formula and, thereby, one has to find the sum of the first n natural numbers. To this end, one can generate several pairs of identical (increasing and decreasing) sequences of natural numbers (Fig. 9.4). Adding two numbers in each row yields the same sum for each pair. By changing the length of the sequence (which is slider-controlled in Fig. 9.4), one can see that this property remains invariant as the length of a string changes. Thus, numerical evidence provided by different strings of natural numbers prompts the summation formula (see formula (2.2), Chapter 2)

$$1+2+3+...+n=\frac{n(n+1)}{2}$$

whence

$$t_n=\frac{n(n+1)}{2}. \qquad (9.4)$$

Relation (9.4) is a closed formula for triangular numbers. Unlike recursive definition (9.2), it allows one to find the triangular number of rank n without knowing the triangular number of rank $n-1$. While both definitions are represented through algebraic formulas, teacher candidates need to know conceptual difference between recursive and closed representations of the sum of the first n counting numbers. This knowledge can be helpful in showing their own students how one can "go from one representation to another, recognize the connections between representations and use different representations appropriately and as needed to solve problems" [Ontario Ministry of Education, 2005b, p. 16]. Thus a useful practice in algebraic thinking could be to derive formula (9.4) from formula (9.2) and vice versa. This would help teacher candidates to understand algebraic ideas behind more and more complex mathematical techniques in order to be able, in turn, to help their own students to "solve ... a variety of routine and non-routine problems with increasing sophistication" [Department for Education, 2013b, p.1].

9.5 Visualizing Mathematical Induction Proof of Formula (9.4)

Formula (9.4) can be proved by the method of mathematical induction (discussed in detail in Chapters 1 and 2) which, as before, can be visually supported. To begin, note that it is true for $n = 1$. Indeed, $t_1 = 1$ and $\frac{1\cdot(1+1)}{2}=1$. The passing from n to $n + 1$ means that for whatever value of n formula is true, it remains true when this value is increased by one. So, let us assume that $2t_n = n(n+1)$ — an equivalent representation of formula (9.4) — is true. We have to show that this assumption implies the relation $2t_{n+1} = (n+1)(n+2)$. As shown in Fig. 9.5, the value $2t_n$ represents the area of rectangle with sides n and $n + 1$. The recursive definition of triangular numbers, that is, relation (9.2), suggests the equality $2t_{n+1} = 2t_n + 2(n+1)$. Figure 9.5 also shows how rectangle with sides n and $n + 1$ can be extended to the rectangle with area $(n+1)(n+2)$ by adding two rectangles each having area $n + 1$. This geometric model provides a visual demonstration of "passing from n to $n + 1$" [Pólya, 1954, p. 111], different from the spreadsheet conditional formatting discussed in Chapter 2. Algebraically, the demonstrative phase can be presented as follows:

$$2t_{n+1} = 2t_n + 2(n+1) = n(n+1) + 2(n+1) = (n+1)(n+2).$$

That is, the relation $t_{n+1} = \dfrac{(n+1)(n+2)}{2}$ confirms that formula (9.4) holds true when n is replaced by $n + 1$. This completes mathematical induction proof of formula (9.4).

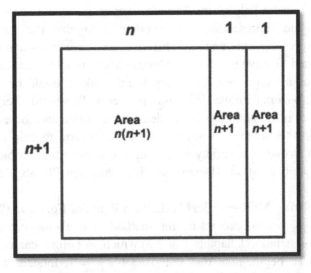

Fig. 9.5. Visual interpretation of passing from n to $n + 1$ in proving formula (9.4).

9.6 Square Test for Triangular Numbers and Its Spreadsheet Implementation

To begin, the notion of the necessary and sufficient condition — one of the major reasoning tools in mathematics — has to be introduced. It is said that condition A is sufficient for condition B if A implies B (i.e., $A \Rightarrow B$). Also, it is said that condition A is necessary for condition B if B implies A (i.e., $B \Rightarrow A$). That is, condition A is necessary and sufficient for condition B if B implies A and vice versa (i.e., $A \Leftrightarrow B$). For example, in order for n to be greater than 5 (condition A) it is necessary that n be greater than zero (condition B). One can see that $A \Rightarrow B$. However, condition B is not sufficient for condition A as a positive number may still be smaller than 5. At the same time, the inequality $n - 5 > 0$ (new condition B) is necessary and sufficient for n to be greater than 5. Indeed, the inequality $n - 5 > 0$ implies $n > 5$ (condition A). That is, because $B \Rightarrow A$, condition B is necessary for condition A. By the same token, the inequality $n > 5$ implies $n - 5 > 0$. That is, because $A \Rightarrow B$, condition B is sufficient for A. Therefore, the inequality $n - 5 > 0$ is necessary and sufficient for the inequality $n > 5$. A more difficult necessary and sufficient condition is represented by

Proposition 9.1. In order for the natural number N to be a triangular number of rank n it is necessary and sufficient that
1) the number $8N + 1$ is a perfect square, and
2) the difference $\sqrt{8N+1} - 1$ is a multiple of two.

Proof. To prove the necessity, let $N = \dfrac{n(n+1)}{2}$ — a triangular number

of rank n. As $n > 0$, the last equality is equivalent to $n = \dfrac{-1 + \sqrt{1+8N}}{2}$

from which the above two conditions follow. To prove the sufficiency, let conditions 1) and 2) be satisfied. Then n is a whole number solution

of the equation $\dfrac{n(n+1)}{2} = N$; that is, N is a triangular number.

	A	B	C	D	E	F	G	H	I	J	K	L	M	N	O	P	Q	R	S	T	U	V
1	1	2	3	4	5	6	7	8	9	10	11	12	13	14	15	16	17	18	19	20	21	22
2	*		*			*				*					*						*	
3																						

Fig. 9.6. Using square test for identifying triangular numbers.

This theorem is called a square test for triangular numbers. The test can be implemented within a spreadsheet using different means, one of which is conditional formatting. Figure 9.6 shows how triangular numbers can be identified among the natural numbers (row 1): either by placing an asterisk underneath (row 2) or through conditional formatting (row 3). To this end, the formula

=IF(AND(SQRT(8*A1+1) = INT(SQRT(8*A1+1)) ,
MOD(SQRT(8*A1+1)-1,2)=0), "*"," ")

can be defined in cell A2 and replicated to the right. In addition, after highlighting the range A3:V3, the formula =A2="*" can be defined in the spreadsheet conditional formatting dialog box (classic style, using a formula to determine which cells to format).

9.7   Developing Square Numbers in Context

Another basic example of a sequence with a variable rate of change is the sequence of square numbers. Like triangular numbers, square numbers can be developed in context using a spreadsheet. To this end, consider

Brain Teaser 9.2: *On the 12th floor of the same hotel another kind of ghost was spotted. It was observed that the ghost also started with room 1, however, unlike the TG, on the second night he emerged in room 4, on the third night he emerged in room 9, then he emerged in room 16, and so on, as shown in Fig. 9.7. Manager of the hotel was seeking answers to new questions, among them:*

What room would the ghost visit on the 7th, 8th, and 9th night?
Would room 200 be a ghost free room? Which room different from room 1 could the ghost visit on the first night in order to end up in room 200?
If this new kind of ghost starts visits with room 1, from which room might the TG start so that both ghosts would visit the same room (not necessarily on the same night)?

9.8   Investigation of Brain Teaser 9.2 Using a Spreadsheet

Figure 9.7 pictures the spreadsheet that generates numbers in row 3 according to the ghost's behavior described in the brainteaser. Once again, one can note that because the ghost always skips an even number of rooms, a room number sought is the sum of the previous such room number and twice the corresponding night number decreased by one. So, one can define in cell B3 the spreadsheet formula =A3+2*B2-1 and replicate it across row 3. As a result, the numbers

$$1, 4, 9, 16, 25, 49, 64, 81, \ldots, 196, 225, \ldots \qquad (9.5)$$

(for which notation s_n will be used below) were generated. This gave the answer to the first two questions of the brainteaser.

B3			f_x	=A3+2*B2-1											
	A	B	C	D	E	F	G	H	I	J	K	L	M	N	O	P
1	☠			☠					☠							☠
2	1	2	3	4	5	6	7	8	9	10	11	12	13	14	15	16
3	1	4	9	16	25	36	49	64	81	100	121	144	169	196	225	256

Fig. 9.7. Square numbers developed in context.

9.9 Developing Formulas for Square Numbers

Recall that the sequence of numbers (9.5) was already encountered through the study of the multiplication table (Chapter 2) and, as the products of two identical factors located on the main diagonal of the table, they were referred to as square numbers. Consequently, this new kind of ghost can be referred to as the square ghost (SG). Also, one can note that the differences between consecutive squares form an arithmetic sequence 3, 5, 7, 9, 11, ... with the difference two, or, more generally, $d_n = 2n - 1$, $n \geq 2$. This observation leads to the following recursive definition of square numbers

$$s_n = s_{n-1} + 2n - 1, \ s_1 = 1. \qquad (9.6)$$

Finally, one can conjecture the closed formula for square numbers

$$s_n = n^2 \qquad (9.7)$$

and prove it by the method of mathematical induction supported by its visual interpretation. To this end, one can note that formula (9.7) is true for $n = 1$. Assuming that formula (9.7) is true for a certain value of n, one has to prove that the relation $s_{n+1} = (n+1)^2$ is also true. Indeed, according to definition (9.6)

$$s_{n+1} = s_n + 2n + 1 = n^2 + 2n + 1 = (n+1)^2.$$

Just as in the case of formula (9.4), the above algebraic proof of formula (9.7) can be given a visual interpretation similar to that of Fig. 9.5. Here, the transition from square with side n to square with side $n + 1$ requires adding two rectangles with sides 1 and n and one unit square; in other words, one has to add $2n + 1$ to n^2 to get $(n + 1)^2$. This completes mathematical induction proof of formula (9.7).

Remark 9.1. The visual interpretation of the transition from n to $n + 1$ in proving formula (9.7) can be seen as a special case of a 'proof without words' of one of the basic algebraic formulas $(n + m)^2 = n^2 + 2nm + m^2$ (when $m = 1$).

9.10   Using Triangular Numbers as Problem-Solving Tools

A new question can be posed in connection with computationally found set of seed values (initial room numbers) that enable the TG to end up in room 100: How can one find those numbers through mathematical reasoning alone in the absence of a spreadsheet? As was already mentioned throughout the book, an important role that technology, in general, and a spreadsheet, in particular, can play as an intellectual tool in one's cognitive development deals with the emergence of residual mental power that can be used in the absence of the tool. The development of such mental power reduces one's dependency on technology and, instead, increases the level of mathematical skills.

To this end, a relationship between the seed values 10, 23, 35, 46, ... (found in Sec. 9.3 by using the spreadsheet of Fig. 9.3 and representing rooms visited by the TG on the first night with the goal to end up in room 100) and triangular numbers can be explored. Without using a spreadsheet one can observe that the seed values, triangular numbers, and the number 100 are connected as follows:

$$10 = 100 - 91 + 1,\ 23 = 100 - 78 + 1,$$

$$35 = 100 - 66 + 1,\ 46\ =\ 100\ -\ 55\ + 1.$$

Furthermore, $23 - 10 = 13$ is the rank of the largest triangular number smaller than 100, $35 - 23 = 13 - 1$, $46 - 35 = 13 - 2$. Numerical evidence suggests that in order to find the room number starting from which the TG can reach room Q, one has to find the rank of largest triangular number not greater than Q. This leads to

Proposition 9.2. Let $R(Q)$ be the rank of the largest triangular number not greater than Q. Then

$$R(Q) = INT(\frac{-1 + \sqrt{1 + 8Q}}{2}). \tag{9.8}$$

Proof. In order to find the rank of the largest triangular number not greater than Q, one has to find the largest positive integer n such that $\frac{n(n+1)}{2} \leq Q$. The latter inequality is equivalent to $n^2 + n - 2Q \leq 0$ which, in turn, implies that $n \leq \frac{-1+\sqrt{1+8Q}}{2}$. The largest integer satisfying the last inequality is equal to $INT(\frac{-1+\sqrt{1+8Q}}{2})$. This completes the proof.

Example. When $Q = 200$ one has $R(200) = INT(\frac{-1+\sqrt{1+1600}}{2}) = 19$, and therefore, the largest triangular number not greater than 200 is the 19th triangular number which is equal to $\frac{19 \cdot 20}{2} = 190$. The 20th triangular number is equal to $\frac{20 \cdot 21}{2} = 210$.

Proposition 9.3. Let $R(Q)$ be the rank of the largest triangular number not greater than Q. Then the sequence of room numbers r_i starting from which the TG can end up in room Q is defined by the recursive formula

$$r_{i+1} = r_i + R(Q) - i + 1 \qquad (9.9)$$

where $r_1 = Q - \frac{(R(Q)+2)(R(Q)-1)}{2}$, $i = 1, 2, ..., R(Q) - 1$.

Proof. One has to prove that formula (9.9) implies the equality $r_{R(Q)} = Q$. To this end, consecutively writing down formula (9.9) for $i = 1, 2, ..., R(Q) - 1$ yields

$$r_2 = r_1 + R(Q)$$
$$r_3 = r_2 + R(Q) - 1$$
$$r_4 = r_3 + R(Q) - 2$$
$$......................$$
$$r_{R(Q)} = r_{R(Q)-1} + R(Q) - (R(Q) - 2).$$

Adding the above $R(Q) - 1$ equalities and cancelling out equal terms results in the chain of equalities

$$r_{R(Q)} = r_1 + R(Q)(R(Q)-1) - (1+2+...+(R(Q)-2))$$

$$= Q - \frac{(R(Q)+2)(R(Q)-1)}{2}$$

$$+R(Q)(R(Q)-1) - \frac{(R(Q)-2)(R(Q)-1)}{2}$$

$$= Q - \frac{R(Q)-1}{2}(R(Q)+2-2R(Q)+R(Q)-2) = Q.$$

This completes the proof.

Remark 9.2. One can computerize the statement of Proposition 9.2 by using a spreadsheet (Fig. 9.8) that integrates formulas (9.8) and (9.9). By changing the content of cell A2, different sequences of seed values can be generated (beginning from cell C2). The spreadsheet of Fig. 9.8 can be programmed as follows.
A2: =100 — given name Q; B2: =INT(0.5*(-1+SQRT(1+8*Q))) — given name R_Q; C1: = 1; D1: =IF(OR(C1=" ",C1=R_Q)," ",C1+1) — replicated across row 1; C2: =A2-0.5*(R_Q+2)*(R_Q-1);
D2: =IF(OR(D1=" ",C1=R_Q)," ",C2+R_Q-C1+1) – replicated across row 2.

	A	B	C	D	E	F	G	H	I	J	K	L	M	N	O
1	Q	R(Q)	1	2	3	4	5	6	7	8	9	10	11	12	13
2	100	13	10	23	35	46	56	65	73	80	86	91	95	98	100

Fig. 9.8. Computerization of Proposition 9.2.

9.11 Two General Formulas for Polygonal Numbers

In general, by setting $P(m, n)$ to denote the n-th polygonal number of side m (where $m \geq 3$, and $P(2, n) = n$), one can derive the recursive formula

$$P(m,n) = P(m,n-1) + (m-2)(n-1) + 1, \; P(m,1) = 1, \qquad (9.10)$$

and the closed formula

$$P(m,n) = 0.5(m-2)(n+1)n - (m-3)n. \qquad (9.11)$$

From formula (9.10), the formulas

$$t_n = t_{n-1} + n, \; s_n = s_{n-1} + 2n - 1, \; p_n = p_{n-1} + 3n - 2, \; h_n = h_{n-1} + 4n - 3,$$

where $t_1 = s_1 = p_1 = h_1 = 1$, result. From formula (9.11), the sequences,

$$t_n = \frac{n(n+1)}{2}, \quad s_n = n^2, \quad p_n = \frac{n(3n-1)}{2}, \quad h_n = n(2n-1) \quad \text{representing,}$$

respectively, triangular ($m = 3$), square ($m = 4$), pentagonal ($m = 5$), and hexagonal ($m = 6$) numbers result. Furthermore, using the identity $\frac{n(n-1)}{2}(m-2) + n = (m-2)\frac{n(n+1)}{2} - (m-3)n$ (which can be verified through *Wolfram Alpha*), formula (9.11) can be rewritten as

$$P(m,n) = \frac{n(n-1)}{2}(m-2) + n. \tag{9.12}$$

Remark 9.3. Formula (9.12) allows one to interpret a polygonal number of rank n as a linear function of its side m with the slope $n(n-1)/2$ — the triangular number of rank $n - 1$. Geometrically, assuming $m \geq 2$, formula (9.12) defines in the plane (m, n) a family of rays (depending on n) with the initial point $(2, n)$ and slope t_{n-1}. Given n, each ray passes through the set of points with integer coordinates such that the difference between the vertical coordinates of two consecutive points is t_{n-1}. Analytically, this can be expressed through yet another formula $P(m,n) = P(m-1,n) + t_{n-1}$ that emphasizes the meaning of a slope as the vertical increment (the rise) when the horizontal increment (the run) is equal to one.

Remark 9.4. Formula (9.12) shows that every polygonal number of an even side is a multiple of its rank and every second polygonal number of an odd side is a multiple of its rank. Indeed, in the former case $m - 2$ is divisible by two; in the latter case, when n is an odd number, $n - 1$ is divisible by two. That is, whereas $P(m,n) = n\left[\frac{n(n-1)(m-2)}{2} + 1\right]$, in both cases the product $n(n-1)(m-2)$ is divisible by two.

Remark 9.5. Formula (9.12) has an interesting connection to the activities on generating cyclic sequences of the form $\{1, 2, 3, \dots, n\}$

described in Sec. 1.9 of Chapter 1. If one divides $P(m, n)$ by $m - 2$, the remainder is equal to $\text{MOD}(n, m - 2)$ — a cycle of the length $m - 2$ with the values $\{1, 2, 3, ..., m - 3, 0\}$, something that can be verified numerically by using a spreadsheet or *Maple*. As shown in the spreadsheet of Fig. 9.9, in column A consecutive natural numbers are defined, in cell D2 the value of m (side of a polygonal number) is defined, in cell B2 the formula =IF(A2=" ", " ", 0.5*A2*(A2-1)*(D\$2-2)+A2) is defined and copied down column B. Finally, in cell C2 the formula

=IF(A2=" ", " ", IF(MOD(B2, D\$2)=0, D\$2-2, MOD(B2, D\$2)))

is defined and replicated down column C. As a result, as shown in Fig. 9.9 for $m = 8$, the cycle $\{1, 2, 3, 4, 5, 6\}$ has been generated.

	A	B	C	D
1	n	$P(m, n)$	cycle	m
2	1	1	1	8
3	2	8	2	
4	3	21	3	
5	4	40	4	
6	5	65	5	
7	6	96	6	
8	7	133	1	
9	8	176	2	
10	9	225	3	
11	10	280	4	
12	11	341	5	
13	12	408	6	
14	13	481	1	
15	14	560	2	
16	15	645	3	
17	16	736	4	
18	17	833	5	
19	18	936	6	
20	19	1045	1	
21	20	1160	2	

Fig. 9.9. Generating the 6-cycle $\{1, 2, 3, 4, 5, 6\}$ via octagonal numbers.

9.12 Square Test for Polygonal Numbers

Proposition 9.4. In order for a natural number N to be a polygonal number of rank n and side m, it is necessary and sufficient that

1) $(m - 4)^2 + 8N(m - 2)$ is a perfect square, and

2) $m - 4 + \sqrt{(m-4)^2 + 8N(m-2)}$ is a multiple of $2(m-2)$.

Proof. To prove the necessity, let $N = \dfrac{n(n-1)}{2}(m-2) + n$ — a polygonal number of rank n and side m. As $n > 0$, the last equality, being a quadratic equation in n, is equivalent to the equation $n = \dfrac{m - 4 + \sqrt{(m-4)^2 + 8N(m-2)}}{2(m-2)}$ from which the above two conditions follow. To prove the sufficiency, let conditions 1) and 2) be satisfied. Then n is a whole number solution of the equation $N = \dfrac{n(n-1)}{2}(m-2) + n$.

9.13 Triangular-Like Numbers

One can introduce what may be called a sequence of triangular-like numbers — triangular numbers uniformly shifted to the right along the number line by a given number. In other words, the TG can start visits from any room — in this case all dangerous room numbers are uniformly shifted triangular numbers. One can develop formulas to calculate room numbers visited by the TG if it starts visits from an arbitrary room a. To this end, substituting the number 1 with a in formula (9.3) yields

$$t_n(a) = a + 2 + 3 + \ldots + n,$$

whence, due to formula (9.4),

$$t_n(a) = \frac{n(n+1)}{2} + a - 1. \tag{9.13}$$

Relation (9.13) is a closed formula for triangular-like numbers $t_n(a)$ as it generates triangular numbers uniformly shifted by $a - 1$ along the number line. For example, when $a = 10$ formula (9.13) yields the sequence $t_n(10) = \{10, 12, 15, 19, 24, 30, 37, 45, \ldots, 100, \ldots\}$. One can see that this sequence includes at least three triangular numbers, namely, 10, 15 and 45. This gives rise to the following question: Given a, how many triangular numbers are there among triangular-like numbers $t_n(a)$? In other words, if two TGs strike the hotel and visit, respectively, rooms 1 and a on the first night, which rooms could potentially be visited by both ghosts and what are the room numbers?

To answer these questions, one has to solve the equation

$$t_n(a) = t_m \qquad (9.14)$$

where $m > n$ and $a > 1$. The constraint $m > n$ (alternatively, $m < n$) indicates that a room visit (if any) by both ghosts may not occur on the same night. Using formulas (9.4) and (9.13), Eq. (9.14) can be replaced by the equation $\dfrac{n(n+1)}{2} + a - 1 = \dfrac{m(m+1)}{2}$ which, in turn, can be simplified to the form

$$(m-n)(m+n+1) = 2(a-1). \qquad (9.15)$$

Equation (9.15) is a two-variable equation with parameter a. Its form suggests that given a, the number of solutions (if any) is finite; that is, within any triangular-like sequence there may only be a finite number of triangular numbers. For example, setting $a = 10$ in Eq. (9.15) yields the equation $(m-n)(m+n+1) = 18$ which, as *Wolfram Alpha* would show, has the whole number solution set $(m,n) = \{(4,1), (5,3), (9,8)\}$, where m and n are, respectively, the ranks of the corresponding triangular number t_m and triangular-like number $t_n(10)$. So, $t_1(10) = 10$ is the 4th triangular number; $t_3(10) = 15$ is the 5th triangular number; and $t_8(10) = 45$ is the 9th triangular number.

In order to further investigate this phenomenon, one can use a spreadsheet to model Eq. (9.15) numerically. Such a spreadsheet would have the form of a two-dimensional table. If both m and n vary within the range $[1, k]$, the straightforward modeling of Eq. (9.15) would require k^2 computations. Therefore, one has to minimize the value of k, given the value of a, or, better still, make the efficient range $[1, k]$ being dependent on a. This investigation will be carried out in the next section.

9.14 Inequalities and Search for Triangular-Like Numbers Using a Spreadsheet

In this section, the machinery of algebraic inequalities described in detail in Chapter 8 will be used in modeling Eq. (9.15) within a spreadsheet. We begin with

Proposition 9.5. The variable m in Eq. (9.15) satisfies the inequality

$$m \leq N \qquad (9.16)$$

where $N = a - 1$.

Proof. Assuming the contrary, that is, assuming $m > N$ yields

$$2N = (m-n)(m+n+1) = m^2 - n^2 + m - n > N^2 - n^2 + N - n$$

whence

$$n^2 + n + N - N^2 > 0. \tag{9.17}$$

When the discriminant D of the quadratic equation $n^2 + n + N - N^2 = 0$ is less than zero, inequality (9.17) is satisfied for all n. However, $D = 1 + 4(N^2 - N) = 4N^2 - 4N + 1 = (2N-1)^2 > 0$ for all $N \neq 1/2$ and therefore inequality (9.17) is not satisfied for all n. A possible solution to (9.17) requires $n > \dfrac{-1 + \sqrt{1 + 4(N^2 - N)}}{2} = \dfrac{-1 + 2N - 1}{2} = N - 1$. In other words, the assumption $m > N$ implies the inequality $n > N - 1$ whence $m + n + 1 > 2N = 2(a - 1)$. The last inequality, however, contradicts to Eq. (9.15) because none of the integer factors may be greater than their product. This completes proof of (9.16).

Corollary 9.1. In Eq. (9.15) the inequality $n \leq N - 1$ holds true.
Proof. Assuming $n > N - 1$ (that is, making an assumption contrary to what has to be proved) yields

$$2N = m^2 - n^2 + m - n < N^2 + N - (N-1)^2 - (N-1) = 2N.$$

That is, we arrived at $2N < 2N$ — a false inequality. Therefore, as the inequality $n \leq N - 1$ cannot be false it must then be true.

Furthermore, one can see that the pair $m = a - 1$, $n = a - 2$ satisfies Eq. (9.15). Indeed,

$$(m-n)(m+n+1) = (a-1-a+2)(2a-3+1) = 2(a-1).$$

Therefore, $(a - 1, a - 2)$ is always a solution to Eq. (9.15) and, thereby, the ranges for m and n can be further improved.

Proposition 9.6. The largest value of m smaller than $a - 1$ that satisfies Eq. (9.15) cannot be greater than $\dfrac{a}{2}$.
Proof. Let the factors in the left-hand side of Eq. (9.15) be as follows: $m - n = 2$, $m + n + 1 = a - 1$. Then their sum $2m + 1 = a + 1$ whence $m = \dfrac{a}{2}$, $n = \dfrac{a}{2} - 2$. Therefore, if a is an even number, then $(\dfrac{a}{2}, \dfrac{a}{2} - 2)$ is a

solution to Eq. (9.15); otherwise, the largest value of m smaller than $a - 1$ that satisfies Eq. (9.15) is smaller than $\dfrac{a}{2}$.

Corollary 9.2. The largest value of n smaller than $a - 2$ that satisfies Eq. (9.15) is not greater than $\dfrac{a}{2} - 2$.

Proof. Because $m \le \dfrac{a}{2}$ in (9.15), it follows that

$$2(a-1) = m^2 - n^2 + m - n \le \frac{a^2}{4} + \frac{a}{2} - n^2 - n$$

whence $n^2 + n - \dfrac{a^2 - 6a + 8}{4} \le 0$. Therefore, $n \le \dfrac{a}{2} - 2$. Indeed, the opposite inequality, $n > \dfrac{a}{2} - 2$, leads to a contradiction:

$$n^2 + n - \frac{a^2 - 6a + 8}{4} > (\frac{a}{2} - 2)^2 + \frac{a}{2} - 2 - \frac{a^2 - 6a + 8}{4}$$

$$= (\frac{a}{2} - 2)(\frac{a}{2} - 1) - \frac{a^2 - 6a + 8}{4} = \frac{(a-4)(a-2) - (a-4)(a-2)}{4} = 0.$$

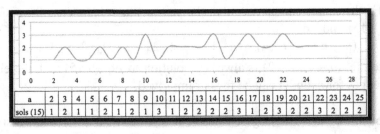

Fig. 9.10. Relating a to the number of triangular numbers among $t_n(a)$ through modeling Eq. (9.15).

In that way, the ranges for the variables m and n can be established, respectively, as $[1, \dfrac{a}{2}]$ and $[1, \dfrac{a}{2} - 2]$. One can use a spreadsheet to construct a function that relates a to the number of solutions of Eq. (9.15). In particular, considering a in the range $[2,100]$ requires the ranges $[1, 50]$ for m and $[1, 48]$ for n. Table and graph representations of the function which relates a to the number of triangular numbers among the terms of the sequence $t_n(a)$ in the range $[1, 25]$ are shown in Fig. 9.10 (assuming that Eq. (9.15) has always a solution $m = a - 1, n = a - 2$).

9.15 Square-Like Numbers

In order to answer the last question posed in Brain Teaser 9.2, one can note that similar to triangular numbers, square numbers can also be shifted along the number line to form what may be called a sequence of square-like numbers. Consider the number 200. The largest square number smaller that 200 is 196. Therefore, shifting the square numbers by four along the number line to the right results in the sequence of square-like numbers $s_n(4) = \{5, 8, 13, 20, ..., 173, 200, ...\}$, which includes 200. The first term of the sequence, $s_1(4) = 1 + (200 - 196) = 1 + 4 = 5$.

Proposition 9.7. Let $R(Q)$ be the rank of the largest square number not greater than Q. Then the sequence of room numbers r_i starting from which the SG can end up in room Q is defined by the recursive formula

$$r_{i+1} = r_i + 2(R(Q) - i + 1) - 1 \qquad (9.18)$$

$$r_1 = Q - R(Q)^2 + 1, \quad i = 1, 2, ..., R(Q) - 1.$$

Proof. One has to prove that formula (9.18) implies the equality $r_{R(Q)} = Q$. To this end, consider the following $R(Q) - 1$ equalities

$$r_2 = r_1 + 2R(Q) - 1$$

$$r_3 = r_2 + 2R(Q) - 3$$

$$r_4 = r_3 + 2R(Q) - 5$$

$$........................$$

$$r_{R(Q)} = r_{R(Q)-1} + 2R(Q) - (2R(Q) - 3).$$

Adding these equalities and cancelling out equal terms yield

$$r_{R(Q)} = r_1 + 2R(Q)(R(Q) - 1) - (1 + 3 + 5 + ... + (2R(Q) - 3))$$

$$= Q - R(Q)^2 + 1 + 2R(Q)(R(Q) - 1) - (R(Q) - 1)^2$$

$$= Q - R(Q)^2 + 1 + 2R(Q)^2 - 2R(Q) - R(Q)^2 + 2R(Q) - 1 = Q.$$

This completes the proof.

One can computerize formula (9.18) by using the spreadsheet shown in Fig. 9.11 that generates the sequence of room numbers starting from which the SG can reach room 200. The spreadsheet of Fig. 9.11 can be programmed as follows.

A2: =200 — given name Q; B2: =INT(SQRT(Q)) — given name R_Q; C1: =1; D1: =IF(OR(C1=" ",C1=R_Q)," ",C1+1) — replicated across row 1; C2: =Q-R_Q^2+1;
D2: =IF(OR(COUNT(D1)=0,C1=R_Q)," ",C2+2*(R_Q-C1+1)-1) — replicated across row 2.

	A	B	C	D	E	F	G	H	I	J	K	L	M	N	O	P
1	Q	R(Q)	1	2	3	4	5	6	7	8	9	10	11	12	13	14
2	200	14	5	32	57	80	101	120	137	152	165	176	185	192	197	200

Fig. 9.11. Computerization of formula (9.18)/Proposition 9.7.

Remark 9.6. In the context of square-like numbers one can explore whether a square-like number can be a square number as well. To this end, note that by shifting square numbers by k yields the equation $s_n + k = s_m$ or $n^2 + k = m^2$. When $k = l^2$, one has the Pythagorean equation

$$n^2 + l^2 = m^2 \qquad (9.19)$$

in which n and m determine the ranks of the coinciding square and square-like numbers. For example, in the case of the classic Pythagorean triple $n = 3$, $l = 4$, $m = 5$, one has $s_3 + 16 = 25 = s_5$. In the case $l = 10$, Eq. (9.19) has the solution $n = 24$, $m = 26$. In general, given l, Eq. (9.19) has at most one solution. Indeed, when $l = 5$, the only integer pair (m, n) satisfying the equation $n^2 + 25 = m^2$ is (13, 12). Otherwise, because $25 = m^2 - n^2 = (m+n)(m-n)$ we would have $m+n = m-n = 5$ whence $n = 0$.

9.16 Two Types of Ghosts on the Same Floor

Consider the situation described in the last question of Brain Teaser 9.2. Let two ghosts, TG and SG, start from different rooms, numbered a and b, respectively, and then move along the same floor with their individual behavioral patterns. How often would both ghosts visit the same room (not necessarily on the same night)? In other words, given a and b, how many triangular-like numbers $t_n(a)$ would coincide with square-like numbers $s_m(b)$? To answer this question, one can construct the equation

$$a - 1 + \frac{n(n+1)}{2} = b - 1 + m^2,$$

transform it to the form

$$m^2 - \frac{n(n+1)}{2} = a - b, \tag{9.20}$$

and then model Eq. (9.20) by using a spreadsheet. Figure 9.12 shows the spreadsheet that calculates positive differences between triangular and square numbers. Consider the case $m = 3$, $n = 2$; Eq. (9.20) turns into the equation

$$a - b = 6 \tag{9.21}$$

which, of course, has infinite number of integer solutions. However, making certain assumptions about the variables a and b, one can narrow down the number of solutions of Eq. (9.21).

If $b < a \leq 10$ in (9.21), then $1 \leq b \leq 4$. One can graph Eq. (9.21) for $1 \leq b \leq 4$ to see that the only lattice points that belong to this graph are $\{(7, 1), (8, 2), (9, 3), (10, 4)\}$. One can further interpret the contextual meaning of, say, the quadruple $(m, n, a, b) = (3, 2, 10, 4)$ that satisfies Eq. (9.20): If TG starts from room 10 and SG starts from room 4 then room 12 will be visited by TG and SG on the 2nd and 3rd nights, respectively. However, as the number 6 appears in the spreadsheet of Fig. 9.12 more than once, there exist many more solutions to Eq. (9.20). This, in turn, raises another question: Is the number of solutions to Eq. (9.20), with the right-hand side equal to six, finite or not? To answer this question, Eq. (9.20) can be rewritten in the form $n^2 + n - 2m^2 + 12 = 0$ whence $n = \dfrac{-1 + \sqrt{8m^2 - 47}}{2}$. For example, when $m = 3$ the last formula gives $n = 2$ — the pair already found. Theoretically, this inquiry leads to the theory of Diophantine equations: finding integer solutions of the equation $8m^2 - 47 = k^2$. This theory is beyond the scope of this book. Nonetheless, one can explore this problem numerically by using a spreadsheet as a one-dimensional modeling tool. Such a spreadsheet would generate eight solutions for the first two thousand values of m. Alternatively, by using *Wolfram Alpha*, one can find closed formulas for

m and k satisfying the last equation. Such use of technology will be discussed in Sec. 9.18 below.

The spreadsheet of Fig. 9.12 can be programmed as follows. A3: = 1; A4: = A3+1 — replicated down column A (hidden from the view); B3: =A3*A4/2 — replicated down column B; C1: =1; D1: = C1 + 1 — replicated across row 1 (hidden from the view); C2: = C1^2 — replicated across row 2; D3: =IF(D\$2>\$B3,D\$2-\$B3," ") — replicated to cell R24.

	B	C	D	E	F	G	H	I	J	K	L	M	N	O	P	Q	R
2		1	4	9	16	25	36	49	64	81	100	121	144	169	196	225	256
3	1		3	8	15	24	35	48	63	80	99	120	143	168	195	224	255
4	3		1	6	13	22	33	46	61	78	97	118	141	166	193	222	253
5	6			3	10	19	30	43	58	75	94	115	138	163	190	219	250
6	10				6	15	26	39	54	71	90	111	134	159	186	215	246
7	15				1	10	21	34	49	66	85	106	129	154	181	210	241
8	21					4	15	28	43	60	79	100	123	148	175	204	235
9	28						8	21	36	53	72	93	116	141	168	197	228
10	36							13	28	45	64	85	108	133	160	189	220
11	45								4	19	36	55	76	99	124	151	180	211
12	55									9	26	45	66	89	114	141	170	201
13	66										15	34	55	78	103	130	159	190
14	78									3	22	43	66	91	118	147	178
15	91										9	30	53	78	105	134	165
16	105											16	39	64	91	120	151
17	120											1	24	49	76	105	136
18	136												8	33	60	89	120
19	153													16	43	72	103
20	171														25	54	85
21	190														6	35	66
22	210															15	46
23	231																25
24	253																3

Fig. 9.12. Modeling Eq. (9.20).

9.17 Exploring m-Gonal-Like Numbers

New context-bounded situations can be posed in order to introduce pentagonal, hexagonal, heptagonal, octagonal, and, more generally, m-gonal numbers through problematic situations involving corresponding ghosts. To this end, consider the first n terms of the arithmetic series with the first term one and difference $m - 2$:

$$1, m - 1, 2m - 3, 3m - 5, \ldots, (n - 1)m - (2n - 3). \qquad (9.22)$$

The sum of these numbers, i.e., the n-th partial sum of the arithmetic series is called the polygonal number of side m and rank n and denoted $P(m, n)$. In order to find this sum, one has to increase n-fold the average of the first and the last terms of sequence (9.22). In doing so, one can arrive at formula (9.11) defining $P(m, n)$.

In order to find $R(Q)$ — the rank of the greatest m-gonal number smaller than or equal to Q, one has to solve the inequality $P(m, n) \leq Q$. Formula (9.11) yields the inequality

$$0.5n(n-1)(m-2) + n \leq Q,$$

which can be simplified to the form

$$(m-2)n^2 + (4-m)n - 2Q \leq 0.$$

The last inequality turns into an equality when

$$n = \frac{m - 4 \pm \sqrt{(m-4)^2 + 8(m-2)Q}}{2(m-2)}.$$

Considering the positive value of n, one has to find

$$R(Q) = FLOOR(\frac{m - 4 + \sqrt{(m-4)^2 + 8(m-2)Q}}{2(m-2)}, 1). \qquad (9.23)$$

Proposition 9.8. Let $R(Q)$ be the rank of the largest m-gonal number smaller than Q satisfying formula (9.23). Then the sequence of room numbers r_i starting from which the m-GG can end up in room Q is defined by the recursive formula

$$r_{i+1} = r_i + (m-2)(R(Q) - i + 1) - (m-3) \qquad (9.24)$$

$$r_1 = Q - 0.5(m-2)R(Q)(R(Q) - 1) - R(Q) + 1, \quad i = 1, 2, ..., R(Q) - 1.$$

Proof. Note that, as above, one has to prove that formula (9.24) implies $r_{R(Q)} = Q$. To this end, writing down formula (9.24) for each value of i one has

$$r_2 = r_1 + (m-2)R(Q) - (m-3)$$

$$r_3 = r_2 + (m-2)(R(Q) - 1) - (m-3)$$

$$r_4 = r_3 + (m-2)(R(Q) - 2) - (m-3)$$

$$\dots\dots\dots\dots\dots\dots\dots\dots\dots$$

$$r_{R(Q)} = r_{r(Q)-1} + (m-2)(R(Q) - (R(Q) - 2)) - (m-3).$$

Adding the above equalities and cancelling out equal terms yield

$$r_{R(Q)} = r_1 + (m-2)R(Q)(R(Q)-1) - (m-2)(1+2+...+(R(Q)-2))$$

$$-(m-3)(R(Q)-1) = Q - 0.5(m-2)R(Q)(R(Q)-1) - R(Q) + 1$$

$$+(m-2)R(Q)(R(Q)-1) - 0.5(m-2)(R(Q)-1)(R(Q)-2)$$

$$-(m-2)(R(Q)-1) + R(Q) - 1$$

$$= Q - (m-2)(R(Q)-1)^2 + (m-2)(R(Q)-1)^2 = Q.$$

This completes the proof.

Finally, Proposition 9.8 can be computerized as shown in the spreadsheet of Fig. 9.13 (where $m = 6$ is entered in cell A2). In row 2 (beginning from cell D2) one can see the sequence of hexagonal-like numbers $\{25, 70, 111, ..., 300\}$ a possible interpretation of which is that a hexagonal ghost starting from room 12 can reach room 300.

	A	B	C	D	E	F	G	H	I	J	K	L	M	N	O
1	m	Q	$R(Q)$	1	2	3	4	5	6	7	8	9	10	11	12
2	6	300	12	25	70	111	148	181	210	235	256	273	286	295	300

Fig. 9.13. Computerization of Proposition 9.8.

The spreadsheet of Fig. 9.13 can be programmed as follows.
A2: = 6 — given name m; B2: =300 — given name Q;
C2: =FLOOR((m-4+SQRT((m-4)^2+8*(m-2)*Q))/(2*(m-2)),1) — given name R_Q; D1: =1; E1: =IF(OR(D1=" ",D1=R_Q)," ",D1+1) — replicated across row 1; D2: =Q-0.5*(m-2)*R_Q*(R_Q-1)-R_Q+1;
E2: =IF(OR(COUNT(E1)=0,D1=Q)," ",D2+(m-2)*(R_Q-D1+1)-(m-3)) — replicated across row 2.

9.18   Two Types of Ghosts Meeting Each Other

Let the two ghosts, TG and SG, meet in room 1 on the first night and then move along the same floor according to their individual behavioral patterns. How often would they visit the same room on the same night? In other words, are there numbers that are both triangular and square? As was mentioned in Chapter 4 (Remark 4.3), the search for squares among triangular numbers is a classic problem of number theory. Such numbers are called triangular squares. It follows from formulas (9.4) and (9.7) that if the triangular number of rank n is also the square number of rank k then $n(n + 1)/2 = k^2$ or

$$n(n+1) = 2k^2. \tag{9.25}$$

That is, triangular squares are defined by equation (9.25). One can attempt to model equation (9.25) using a one-dimensional spreadsheet. In doing so, five triangular squares can be found among the first two thousand triangular numbers, including two obvious numbers, 1 and 36. Alternatively, one can enter into the input box of *Wolfram Alpha* the command "solve over the integers n(n+1)=2k^2, n>0, k>0" to get the formula

$$k = \frac{(3+2\sqrt{2})^m - (3-2\sqrt{2})^m}{4\sqrt{2}}, m = 0, 1, 2, \dots, \tag{9.26}$$

which appears in [Euler, 1849, p. 267]. Noting that $3 \pm 2\sqrt{2} = (1 \pm \sqrt{2})^2$, TS_m — the triangular square of rank m — can be found by modifying formula (9.26) to the form

$$TS_m = \left[\frac{(1+\sqrt{2})^{2m} - (1-\sqrt{2})^{2m}}{4\sqrt{2}} \right]^2, m = 0, 1, 2, \dots. \tag{9.27}$$

Formula (9.27), which is a rather obvious modification of formula (9.26), nonetheless, without any reference to Euler, was given by Roberts [1879, p. 17] as a problem to demonstrate "that triangular numbers which are also squares are given [by formula (9.27)]" (see also [Dickson, 1971, p. 27]). Note that formulas (9.26) and (9.27) are similar to formulas (10.27) and (10.28) for Fibonacci and Lucas numbers, respectively, discussed in the next chapter. Triangular squares, however, grow very fast and already the fifth triangular square is a number with seven digits (for comparison, the fifth Fibonacci and Lucas numbers are 5 and 7, respectively). A spreadsheet, being limited in accuracy to displaying numbers with not more than 15 digits, can only handle formula (9.27) for $n \in [0, 10]$ (Fig. 9.14, column B).

Wolfram Alpha can generate larger triangular squares using formula (9.26) if one enters "Table [$((3+2\sqrt{2})^m - (3-2\sqrt{2})^m)^2 / 32$, {m, 20}]" into its input box yielding the result shown in Fig. 9.15. In particular, the 20th triangular square is a 30-digit number. Furthermore, as shown in Fig. 9.16, *Maple* can also generate triangular squares using formula (9.27) and the command seq(simplify(TS(m), m=1..20), through which different ways of computing can be compared to observe same results. Finally, one can discover that the ratios of two consecutive triangular squares converge to a number close to 34 (Fig. 9.14, column C; Fig. 9.16, the bottom part). The exact value of this number, which is similar to the Golden Ratio (Sec. 10.4 of Chapter 10), can be found as follows (noting that $(\sqrt{2}+1)(\sqrt{2}-1)=1$):

$$\lim_{m\to\infty}\frac{TS_{m+1}}{TS_m} = \lim_{m\to\infty}\left[\frac{(\sqrt{2}+1)^{2(m+1)} - (\sqrt{2}-1)^{2(m+1)}}{(\sqrt{2}+1)^{2m} - (\sqrt{2}-1)^{2m}}\right]^2$$

$$= \lim_{m\to\infty}\left[\frac{1}{(\sqrt{2}+1)^2} \cdot \frac{(\sqrt{2}+1)^{4(m+1)} - 1}{(\sqrt{2}+1)^{4m} - 1}\right]^2$$

$$= \lim_{m\to\infty}\left[\frac{1}{(\sqrt{2}+1)^2} \cdot \frac{(\sqrt{2}+1)^4 - \dfrac{1}{(\sqrt{2}+1)^{4m}}}{1 - \dfrac{1}{(\sqrt{2}+1)^{4m}}}\right]^2$$

$$= (\sqrt{2}+1)^4 = (3+2\sqrt{2})^2 = 17 + 12\sqrt{2}.$$

	A	B	C
1	1	1	
2	2	36	36
3	3	1225	34.02777778
4	4	41616	33.9722449
5	5	1413721	33.97061226
6	6	48024900	33.97056421
7	7	1631432881	33.97056279
8	8	55420693056	33.97056275
9	9	1882672131025	33.97056275
10	10	63955431761796	33.97056275

Fig. 9.14. Triangular squares and their ratios.

Alternate form:

{1, 36, 1225, 41 616, 1 413 721, 48 024 900, 1 631 432 881,
55 420 693 056, 1 882 672 131 025, 63 955 431 761 796,
2 172 602 007 770 041, 73 804 512 832 419 600,
2 507 180 834 294 496 361, 85 170 343 853 180 456 676,
2 893 284 510 173 841 030 625, 98 286 503 002 057 414 584 576,
3 338 847 817 559 778 254 844 961, 113 422 539 294 030 403 250 144 100,
3 853 027 488 179 473 932 250 054 441,
130 889 512 058 808 083 293 251 706 896}

Fig. 9.15. Computing triangular squares using *Wolfram Alpha*.

$$\texttt{> } TS(m) := \left(\frac{\left(1+\sqrt{2}\right)^{2m} - \left(1-\sqrt{2}\right)^{2m}}{4\sqrt{2}} \right)^2$$

$$TS := m \to \frac{1}{16} \frac{\left(\left(1+\sqrt{2}\right)^{2m} - \left(1-\sqrt{2}\right)^{2m}\right)^2}{\left(\sqrt{2}\right)^2}$$

>
> seq(simplify(TS(m)), m = 1..20)

1, 36, 1225, 41616, 1413721, 48024900, 1631432881, 55420693056, 1882672131025, 63955431761796, 2172602007770041, 73804512832419600, 2507180834294496361, 8517034385 3180456676, 2893284510173841030625, 98286503002057414584576, 3338847817559778254844961, 113422539294030403250144100, 3853027488179473932250054441, 130889512058808083293251706896

$$\texttt{> } seq\left(evalf\left(\frac{TS(m+1)}{TS(m)} \right), m = 1..20 \right)$$

35.99999999, 34.02777776, 33.97224484, 33.97061226, 33.97056418, 33.97056278, 33.97056274, 33.97056270, 33.97056274, 33.97056272, 33.97056271, 33.97056274, 33.97056273, 33.97056272, 33.97056273, 33.97056275, 33.97056271, 33.97056274, 33.97056273, 33.97056275

Fig. 9.16. Computing triangular squares and their ratios using *Maple*.

Chapter 10

Advanced Explorations with Fibonacci Numbers

10.1 Introduction

In the previous chapters, a spreadsheet was used to model various numeric sequences. They included arithmetic sequences, cyclic sequences, sums of numbers having different properties within the addition and multiplication tables, prime numbers, polygonal and polygonal-like numbers. Some of those numbers were defined by a formula; other numbers were defined without using formula (e.g., cyclic sequences or prime numbers). In this chapter, a famous number sequence of a different kind will be considered. Its properties are quite distinct from the properties of numbers just mentioned. These new numbers will also be used as a basis for generating new number sequences, all in the context of integrated spreadsheets.

To begin, consider the number sequence

$$1, 1, 2, 3, 5, 8, 13, 21, 34... \qquad (10.1)$$

in which every number beginning from the third is the sum of the two preceding numbers and the first two numbers are equal to one. As is well known, these numbers are named after Fibonacci[p], who introduced them into the Western mathematics by exploring the growth of the population of rabbits breeding in ideal circumstances. Sequence (10.1), however, was known long before Fibonacci appearing in the seventh century Indian mathematics [Singh, 1985]. The importance of sequence (10.1) for mathematical applications is, in particular, due to the fact it "can be found in many natural patterns like in pineapples, sunflowers, nautilus and pine cones" [Science Centre Singapore, 2010] and can be used "in art, music, and architecture" [BrainPop UK, 2014]. The sequence of Fibonacci numbers has been mentioned as a special example of a function to remind high school students "that sequences are functions,

---

[p] Leonardo Fibonacci — the most prominent Italian mathematician of the twelfth-thirteenth centuries, credited with the introduction of Hindu-Arabic number system into the Western world.

sometimes defined recursively, whose domain is a subset of the integers" [Common Core State Standards, 2010, p. 69].

This chapter will demonstrate that Fibonacci numbers and associated concepts when explored using integrated spreadsheets provide an excellent context for promoting many ideas of mathematics education reform among teacher candidates allowing them to "take actions like representing, experimenting, modeling, classifying, and proving" (Conference Board of the Mathematical Sciences, 2001, p. 8). This context can be included in the recommended discrete mathematics and computer science courses for the candidates for it supports "topics such as ... finite difference equations, iterations and recursion; ... and computer programming" (Conference Board of the Mathematical Sciences, 2012, p. 66).

One can define Fibonacci numbers F_n through the following difference equation

$$F_{n+1} = F_n + F_{n-1} \,, \tag{10.2}$$

subject to initial conditions

$$F_0 = F_1 = 1 \,. \tag{10.3}$$

Changing one of the first two Fibonacci numbers (or both) and keeping its (recursive) definition without change, yields a new number sequence that is commonly referred to as a Fibonacci-like sequence. The most famous example of that kind is the sequence of Lucas[q] numbers

$$2, 1, 3, 4, 7, 11, 18, 29, 47, \dots . \tag{10.4}$$

Coincidentally, it was Lucas [1891] who gave sequence (10.1) its name [Koshy, 2001]. Lucas numbers can be defined similarly to Fibonacci numbers as follows

$$L_{n+1} = L_n + L_{n-1} \,, \tag{10.5}$$
$$L_0 = 2, L_1 = 1 \,. \tag{10.6}$$

The recursive nature of Fibonacci and Lucas numbers is often used as the demonstration of the spreadsheet facility to generate numbers through the ostensive definition: given the first two terms, definition (10.2) (or (10.5)) is communicated to the spreadsheet by pointing at the givens. The spreadsheet that generates Fibonacci numbers is shown in

---

[q] Edouard Lucas — a French mathematician of the nineteenth century, known also as the inventor of the Tower of Hanoi puzzle [Stockmeyer *et al.*, 1995].

Fig. 10.1 where the formula bar includes definition (10.2) formulated in the language of cell references. By attaching sliders to the cells containing the first two terms, different Fibonacci-like sequences can be generated (e.g., Lucas numbers in Fig. 10.2). Fibonacci-like sequences are the most natural extensions of the classic sequence (10.1). In the context of spreadsheets, proceeding from Fibonacci numbers, one can learn using the tool for numerical modeling of difference equations, and, in doing so, to motivate the development of abstract mathematical concepts utilizing the power of numerical examples.

	C1		⊗ ⊘	*fx*	=B1+A1									
	A	B	C	D	E	F	G	H	I	J	K	L	M	N	O
1	1	1	2	3	5	8	13	21	34	55	89	144	233	377	610
2															

Fig. 10.1. Fibonacci numbers.

	C1		⊗ ⊘	*fx*	=B1+A1									
	A	B	C	D	E	F	G	H	I	J	K	L	M	N	O
1	2	1	3	4	7	11	18	29	47	76	123	199	322	521	843
2															
3															
4															

Fig. 10.2. Lucas numbers.

10.2 Extending Fibonacci Numbers to New Contexts

One can extend the context of Fibonacci numbers by changing the rule according to which the numbers develop. For example, consider the sequence

$$1, 1, 3, 7, 17, 41, \ldots \qquad (10.7)$$

in which every number beginning from the third is twice the previous number plus the one that precedes it. Just as number sequences (10.1) and (10.4) can be formally defined by Eqs. (10.2)-(10.3) and (10.5)-(10.6), sequence (10.7) satisfies the difference equation

$$f_{n+1} = 2f_n + f_{n-1} \qquad (10.8)$$

with the initial conditions

$$f_0 = f_1 = 1. \qquad (10.9)$$

The coefficient in f_n can be considered a parameter, and in the spreadsheet environment of Fig. 10.3 a slider-controlled cell B2 can be given the name "a" so that sequence (10.7) and other number sequences (this time, we do not call them Fibonacci-like numbers because definition (10.2) does not hold true) can be generated through the formula shown in the formula bar (Fig. 10.3).

C4			fx	=a*B4+A4					
	A	B	C	D	E	F	G	H	I	J
1										
2		2								
3										
4	1	1	3	7	17	41	99	239	577	1393
5										

Fig. 10.3. Modeling Eq. (10.7).

Another avenue which extends the exploration of Fibonacci numbers through the use of a spreadsheet and leads to parameterization is as follows. Consider the sequence

$$1, 2, 5, 13, 34, 89, \ldots \qquad (10.10)$$

obtained by eliminating every second term in sequence (10.1). One can check to see that sequence (10.10) satisfies the difference equation

$$f_{n+1} = 3f_n - f_{n-1} \qquad (10.11)$$

where

$$f_0 = 1, f_1 = 2. \qquad (10.12)$$

By using the spreadsheet shown in Fig. 10.4, one can discover that, regardless of initial values, by eliminating every second term from any Fibonacci-like sequence one can always obtain the sequence of numbers described by difference equation (10.11). For example, eliminating every second term from the sequence 2, 4, 6, 10, 16, 26, 42, 68, 110, ... yields the sequence 2, 6, 16, 42, 110, ... in which $16 = 3 \cdot 6 - 2$, $42 = 3 \cdot 16 - 6$, $110 = 3 \cdot 42 - 16$; that is, the resulting sequence appears satisfying Eq. (10.11). This is quite an unexpected phenomenon: by taking the first and the third terms of any Fibonacci-like sequence as the initial values for a new sequence developed through Eq. (10.11), one eliminates every second term of the original sequence. Difference equation (10.11) can be referred to as Fibonacci sieve of order one.

C6			fx	=a*B6−A6								
	A	B	C	D	E	F	G	H	I	J	K	L	M
1													
2		3											
3													
4	1	1	2	3	5	8	13	21	34	55	89	144	233
5													
6	1	2	5	13	34	89	233	610	1597	4181	10946	28657	75025
7													
8													
9	2	7	9	16	25	41	66	107	173	280	453	733	1186
10													
11	2	9	25	66	173	453	1186	3105	8129	21282	55717	145869	381890

Fig. 10.4. Rows 6 and 11, respectively, generate every 2nd number of rows 4 and 9.

C4		fx	=a*C3+b*C2
	A	B	C	D
1	0	a	f_n	
2	sign slider	3	2	
3			1	
4			5	
5			16	
6	3		53	
7			175	
8	value slider		578	
9	0	b	1909	
10	sign slider	1	6305	
11			20824	
12			68777	
13			227155	
14	1		750242	
15			2477881	
16	value slider		8183885	
17			27029536	

Fig. 10.5. Modeling Eq. (10.13) for $a = 3, b = 1, \alpha = 2, \beta = 1$.

Now, one can consider another generalization of Fibonacci numbers by introducing parameters a, b, α, and β into Eqs. (10.2)-(10.3) as follows

$$f_{n+1} = af_n + bf_{n-1}, \tag{10.13}$$

$$f_0 = \alpha,\ f_1 = \beta. \tag{10.14}$$

In the next section, it will be demonstrated how the introduction of such a generalization can be motivated by the development of Fibonacci sieves of order greater than one.

The spreadsheet used to model Eqs. (10.13)-(10.14) for different values of the parameters is shown in Fig. 10.5. It displays the particular case of Eqs. (10.11)-(10.12), i.e., the case $a = 3$, $b = 1$, $\alpha = 2$, $\beta = 1$. Note that in order to allow for both positive and negative integer values of parameters a and b, two kinds of sliders have to be used: a value slider and a sign slider. The sign slider is assigned two numerical values only: 0 as the minimum value and 1 as the maximum value. In the former case, the value slider (that can only be positive) is kept without change; in the latter case, the value slider is negated. As the variation of the initial values does not qualitatively alter the behavior of sequence (10.13), the change of parameters α (cell C2) and β (cell C3) can be manually controlled.

10.3   Fibonacci Sieve of Order k

One may wonder as to what the values of parameters a and b are in the case of Fibonacci sieve of order two; that is, when the process of elimination of every second number, this time, from sequence (10.10), is repeated? To answer this question, one has to find a difference equation that generates the subsequence 1, 5, 34, 233, ... of Fibonacci numbers. One may first note that the three numbers 1, 5, and 34 can be linked through the equality $34 = 7 \cdot 5 - 1$, which has the form of Eq. (10.11), and then check to see that the numbers 5, 34, and 233 are similarly connected. Indeed, $233 = 7 \cdot 34 - 5$. Then, the next term is $7 \cdot 233 - 34 = 1597$ and the number 1597 coincides with the 16th Fibonacci number. That is, in the case of Fibonacci sieve of order two we have $a = 7$ and $b = -1$ (see Eq. (10.20) below).

The next step is to eliminate every second number from the sequence 1, 5, 34, 233, 1597, ... , and guess an equation that describes the resulting number sequence 1, 34, 1597, ... In order to have a similar link among the first three terms of the last sequence, one can solve the equation $1597 = a \cdot 34 - 1$ to get $a = 47$. That is, in the case of Fibonacci sieve of order three we have $a = 47$ and $b = -1$ (see Eq. (10.23) below).

To describe our findings in a more general form, let us consider the sequence of Fibonacci numbers

$$F_0, F_1, F_2, \ldots, F_n, \ldots \qquad (10.15)$$

and eliminate every second number from (10.15) to get a new sequence

$$F_0, F_2, F_4, \ldots, F_{2n}, \ldots \qquad (10.16)$$

which, as was shown above, satisfies the equations

$$f_{n+1} = 3f_n - f_{n-1} \qquad (10.17)$$

$$f_0 = F_0, f_1 = F_2. \qquad (10.18)$$

The second step was to eliminate every second number from (10.16) to get

$$F_0, F_4, F_8, \ldots, F_{4n}, \ldots . \qquad (10.19)$$

As was conjectured above, sequence (10.19) satisfies the equations

$$f_{n+1} = 7f_n - f_{n-1} \qquad (10.20)$$

$$f_0 = F_0, f_1 = F_4. \qquad (10.21)$$

Difference equation (10.20) with initial data (10.21) can be referred to as Fibonacci sieve of order two.

On the third step, eliminating every second number from sequence (10.19) yields the sequence

$$F_0, F_8, F_{16}, \ldots, F_{8n}, \ldots . \qquad (10.22)$$

Our conjecture was that sequence (10.22) satisfies the equations

$$f_{n+1} = 47f_n - f_{n-1} \qquad (10.23)$$

$$f_0 = F_0, f_1 = F_8. \qquad (10.24)$$

Difference equation (10.23) with initial data (10.24) can be referred to as Fibonacci sieve of order three.

Do the coefficients 3, 7, and 47 in recurrences (10.17), (10.20), and (10.23) form any pattern? One can look at the spreadsheet of Fig. 10.2 where these numbers appear in cells C1, E1, and I1. It turns out that the parameters of Fibonacci sieves of order one, two, and three, are Lucas numbers L_2, L_4, and L_8, respectively. One can further conjecture that the next Lucas number in this sequence is $L_{16} = 2207$. This conjecture can be verified numerically by using the spreadsheet of Fig. 10.6 (column G); cell G5 displays Fibonacci number $F_{32} = 3524578$ — the smallest number that survives Fibonacci sieve of order five. Here are programming details of the spreadsheet-based sieve shown in Fig. 10.6.

The range of rows 2, 3, and 4 is hidden from view. (C2): =1, (D2): =1+C2 — replicated to cell H2, (C3): =2^C2 — replicated to cell H3, (C4): =2^(C2-1) — replicated to cell H4, (C5): = SMALL(C6:C104, 2) — replicated to cell H5. The last formula collects the second smallest number surviving Fibonacci sieve of order one (column C), two (column D), ..., six (column H). (C6): =$B6 — replicated to cell H6. (C7): =IF(MOD($A7, C$3)=0, B7, " ") — replicated across columns and down rows to cell H105.

	A	B	C	D	E	F	G	H
1								
5	n	$F_n\backslash f_{1,k}$	2	5	34	1597	3524578	17167680177565
6	0	1	1	1	1	1	1	1
7	1	1						
8	2	2	2					
9	3	3						
10	4	5	5	5				
11	5	8						
12	6	13	13					
13	7	21						
14	8	34	34	34	34			
15	9	55						
16	10	89	89					
17	11	144						
18	12	233	233	233				
19	13	377						
20	14	610	610					
21	15	987						
22	16	1597	1597	1597	1597	1597		
23	17	2584						
24	18	4181	4181					
25	19	6765						
26	20	10946	10946	10946				
27	21	17711						
28	22	28657	28657					
29	23	46368						
30	24	75025	75025	75025	75025			

Fig. 10.6. Fibonacci sieves of order one through six.

Moving toward generalization, one has to find a difference equation, alternatively, Fibonacci sieve of order k, that generates a sub-sequence F_{n_i} of Fibonacci numbers after the process of the elimination of every second number is applied to sequence (10.1) k times. To this end, one can observe that the numbers 2, 4, 8, and 16 (the ranks of the above four Lucas numbers) are the powers of two. Generalizing from the special cases leads to

Proposition 10.1. Let F_i and L_i be, respectively, Fibonacci and Lucas numbers of rank i. Then the family of difference equations

$$f_{n+1,k} = L_{2^k} f_{n,k} - f_{n-1,k} \qquad (10.25)$$

subject to the initial conditions

$$f_{0,k} = F_0, f_{1,k} = F_{2^k}, \qquad (10.26)$$

where $k = 1, 2, 3, \ldots$, and $f_{n,k} = F_{n \cdot 2^k}$, describe Fibonacci sieve of order k.

This proposition, motivated by spreadsheet modeling, can be proved using the following closed formulas for Fibonacci and Lucas numbers, respectively,

$$F_n = \frac{1}{\sqrt{5}} \left[\left(\frac{1+\sqrt{5}}{2} \right)^{n+1} - \left(\frac{1-\sqrt{5}}{2} \right)^{n+1} \right] \qquad (10.27)$$

and

$$L_n = \left(\frac{1+\sqrt{5}}{2} \right)^n + \left(\frac{1-\sqrt{5}}{2} \right)^n \qquad (10.28)$$

known as Binet's formulas[r]. The derivation of formulas (10.27) and (10.28) is beyond the scope of this book and, along with the proof of Proposition 1 can be found in [Abramovich and Leonov, 2009].

Remark 10.1. Using *Maple*, one can obtain formulas (10.27) and (10.28) by means of the operator *rsolve* (Fig. 10.7); some ·uncomplicated transformations are required to simplify *SolFib* to the form of (10.27). Likewise, a simple query, containing the words "Fibonacci numbers formula", may be entered into the input box of *Wolfram Alpha* to get a hint towards the creation of formula (10.27) in terms of the Golden Ratio ϕ and then getting the latter both in approximate and exact forms (Fig. 10.8). Furthermore, computations shown in Fig. 10.6 can be extended to

---

[r] Jacques Philippe Marie Binet — a French mathematician of the nineteenth century.

compute Fibonacci and Lucas numbers (as well as their subsequences with indices being the powers of two) the number of digits of which far exceeds the capacity of a spreadsheet that, as was mentioned in Chapter 7, makes it possible to display with precision a number with at most 15 digits. So, integrating a spreadsheet with *Maple* makes it possible not only to support extended symbolic computations but numeric computations as well. Such computations are shown in Fig. 10.9.

$$Eq1 := f(n+1) = f(n) + f(n-1);$$
$$f(n+1) = f(n) + f(n-1)$$

$$Fib := f(0) = 1, f(1) = 1;$$
$$f(0) = 1, f(1) = 1$$

$$Luc := f(0) = 2, f(1) = 1;$$
$$f(0) = 2, f(1) = 1$$

$$SolFib := rsolve(\{Eq1, Fib\}, f)$$
$$\left(\frac{1}{10}\sqrt{5} + \frac{1}{2}\right)\left(\frac{1}{2}\sqrt{5} + \frac{1}{2}\right)^n + \left(\frac{1}{2} - \frac{1}{10}\sqrt{5}\right)\left(-\frac{1}{2}\sqrt{5} + \frac{1}{2}\right)^n$$

$$SolLuc := rsolve(\{Eq1, Luc\}, f)$$
$$\left(\frac{1}{2}\sqrt{5} + \frac{1}{2}\right)^n + \left(-\frac{1}{2}\sqrt{5} + \frac{1}{2}\right)^n$$

Fig. 10.7. Using *Maple* in solving recurrences (10.2)-(10.3) and (10.5)-(10.6).

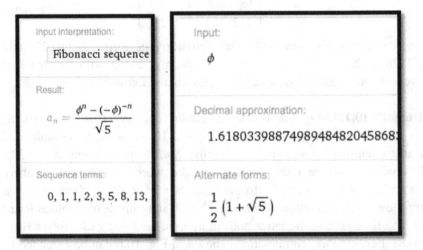

Fig. 10.8. Using *Wolfram Alpha* in developing formula (10.27).

$$> F(n) := \frac{1}{\sqrt{5}}\left(\left(\frac{1+\sqrt{5}}{2}\right)^{n+1} - \left(\frac{1-\sqrt{5}}{2}\right)^{n+1}\right)$$

$$F := n \rightarrow \frac{\left(\frac{1}{2}+\frac{1}{2}\sqrt{5}\right)^{n+1} - \left(\frac{1}{2}-\frac{1}{2}\sqrt{5}\right)^{n+1}}{\sqrt{5}}$$

$$> L(n) := \left(\left(\frac{1+\sqrt{5}}{2}\right)^{n} + \left(\frac{1-\sqrt{5}}{2}\right)^{n}\right)$$

$$L := n \rightarrow \left(\frac{1}{2}+\frac{1}{2}\sqrt{5}\right)^{n} + \left(\frac{1}{2}-\frac{1}{2}\sqrt{5}\right)^{n}$$

> $seq(simplify(F(n)), n = 91..95)$
7540113804746346429, 12200160415121876738, 19740274219868223167, 31940434634990099905, 51680708854858323072
> $seq(simplify(L(n)), n = 91..95)$
10420180999117162549, 16860207025497407047, 27280388024614569596, 44140595050111976643, 71420983074726546239

Fig. 10.9. Computing large Fibonacci and Lucas numbers using *Maple*.

10.4 From the Golden Ratio to Its Generalization

A spreadsheet can be further used to explore the behavior of the ratios F_{n+1}/F_n and L_{n+1}/L_n generated by difference equations (10.2) and (10.6), respectively (Fig. 10.10, columns G and H). This simple exploration will demonstrate that regardless of the initial values of a Fibonacci-like sequence, as n grows larger, the ratios of two consecutive terms of the sequence approach the same number, $\frac{1+\sqrt{5}}{2}$ $(=1.6180334...)$, called the Golden Ratio. In much the same way, the behavior of the ratios f_{n+1}/f_n generated by Eq. (10.13) can be explored for different values of a and b. This exploration can be interpreted as yet another extension of the context of Fibonacci numbers leading to the notion of a generalized golden ratio.

In order to explain the phenomenon of convergence of the ratios to the Golden Ratio regardless of initial values (10.14) of Fibonacci-like sequence (10.13), consider the sequence

$$x, y, x+y, x+2y, 2x+3y, 3x+5y, 5x+8y, 8x+13y, 13x+21y,... \quad (10.29)$$

which develops from (10.13)-(10.14) when $a = b = 1$, $f_0 = x$, $f_1 = y$. Setting $f_8(x, y)$ to be the 8th term of sequence (10.29), one can note that the coefficients in x and y are, respectively, the 6th and 7th Fibonacci numbers. Therefore, $f_8(x,y) = F_6 \cdot x + F_7 \cdot y$.

In general, using the method of mathematical induction, one can show that

$$f_n(x,y) = F_{n-2} \cdot x + F_{n-1} \cdot y \qquad (10.30)$$

is the n-th term of sequence (10.29), $n = 0, 1, 2, \ldots$. To this end, noting that $f_2(x,y) = x + y = F_0 \cdot x + F_1 \cdot y$ to confirm the base clause, and assuming that both (10.30) and the relation $f_{n-1}(x,y) = F_{n-3} \cdot x + F_{n-2} \cdot y$ hold true, one can write

$$f_{n+1}(x,y) = f_{n-1}(x,y) + f_n(x,y) = F_{n-3} \cdot x + F_{n-2} \cdot y + F_{n-2} \cdot x + F_{n-1} \cdot y$$
$$= (F_{n-3} + F_{n-2}) \cdot x + (F_{n-2} + F_{n-1}) \cdot y = F_{n-1} \cdot x + F_n \cdot y,$$

thus completing mathematical induction proof of relation (10.30).

Due to the relation $\lim\limits_{n \to \infty} \dfrac{F_n}{F_{n-1}} = \dfrac{1 + \sqrt{5}}{2}$, the ratios of two consecutive terms of sequence (10.29) tend to the Golden Ratio as well. Indeed,

$$\frac{f_{n+1}(x,y)}{f_n(x,y)} = \frac{F_{n-1} \cdot x + F_n \cdot y}{F_{n-2} \cdot x + F_{n-1} \cdot y} = \frac{F_{n-1} \cdot x(1 + \dfrac{F_n}{F_{n-1}} \cdot \dfrac{y}{x})}{F_{n-1} \cdot x(\dfrac{F_{n-2}}{F_{n-1}} + \dfrac{y}{x})} = \frac{1 + \dfrac{F_n}{F_{n-1}} \cdot \dfrac{y}{x}}{\dfrac{y}{x} + \dfrac{1}{\dfrac{F_{n-1}}{F_{n-2}}}}$$

$$\xrightarrow[n \to \infty]{} \frac{1 + \dfrac{1 + \sqrt{5}}{2} \cdot \dfrac{y}{x}}{\dfrac{y}{x} + \dfrac{2}{1 + \sqrt{5}}} = \frac{1 + \sqrt{5}}{2} \cdot \frac{\dfrac{2}{1 + \sqrt{5}} + \dfrac{y}{x}}{\dfrac{y}{x} + \dfrac{2}{1 + \sqrt{5}}} = \frac{1 + \sqrt{5}}{2}.$$

This explains that not only the ratios of two consecutive Lucas numbers converge to the Golden Ratio, but also the ratios of two consecutive terms of any Fibonacci-like sequence converge to the Golden Ratio. Put another way, the Golden Ratio turns out to be an invariant for the whole family of Fibonacci-like sequences.

In much the same way, one can explore the behavior of the ratios of two consecutive terms of subsequences of Fibonacci-like sequences such as those surviving Fibonacci sieves of order $k \geq 1$. As it turns out, such ratios do not tend to the Golden Ratio and, therefore, the numbers which they approach as n tends to infinity will be called below generalized golden ratios. Some other generalizations of the Golden Ratio are discussed in [Rakocevic, 2004; Stakhov, 2004; Abramovich and Leonov, 2009].

10.5 Generalized Golden Ratios

Assuming that $\lim\limits_{n\to\infty}\dfrac{f_{n+1,k}}{f_{n,k}}$ exists and is equal to $l(k)$, it follows from relation (10.25) that

$$\frac{f_{n+1,k}}{f_{n,k}} = L_{2^k} - \frac{1}{f_{n,k} / f_{n-1,k}}$$

and, therefore, $l(k) = L_{2^k} - \dfrac{1}{l(k)}$, whence

$$l(k) = \frac{L_{2^k} + \sqrt{L_{2^k}^{\ 2} - 4}}{2} . \tag{10.31}$$

Now, using formula (10.31), one can show that unlike the case when $\lim\limits_{n\to\infty}\dfrac{F_{n+1}}{F_n} = \dfrac{1+\sqrt{5}}{2}$, the sequence $l(k) = \lim\limits_{n\to\infty}\dfrac{f_{n+1,k}}{f_{n,k}}$ increases monotonically as the value of k (i.e., the order of a Fibonacci sieve) increases. Indeed,

$$l(k+1) - l(k) = \frac{L_{2^{k+1}} + \sqrt{L_{2^{k+1}}^{\ 2} - 4}}{2} - \frac{L_{2^k} + \sqrt{L_{2^k}^{\ 2} - 4}}{2} = \frac{L_{2^{k+1}} - L_{2^k}}{2}$$

$$+ \frac{L_{2^{k+1}} - L_{2^k}}{2(\sqrt{L_{2^{k+1}}^{\ 2} - 4} + \sqrt{L_{2^k}^{\ 2} - 4})} = \frac{L_{2^{k+1}} - L_{2^k}}{2}(1 + \frac{1}{\sqrt{L_{2^{k+1}}^{\ 2} - 4} + \sqrt{L_{2^k}^{\ 2} - 4}}) > 0$$

for $k = 2, 3, 4, \ldots$.

Furthermore,

$$\mid \frac{L_{2^k} + \sqrt{L_{2^k}^{\,2} - 4}}{2} - L_{2^k} \mid = \frac{\mid L_{2^k} - \sqrt{L_{2^k}^{\,2} - 4} \mid}{2} = \frac{L_{2^k}^{\,2} - L_{2^k}^{\,2} + 4}{2(L_{2^k} + \sqrt{L_{2^k}^{\,2} - 4})}$$

$$= \frac{2}{L_{2^k} + \sqrt{L_{2^k}^{\,2} - 4}} \xrightarrow{k \to \infty} 0.$$

As shown in the spreadsheet of Fig. 10.11, already for Fibonacci sieve of order three we have $\mid \lim\limits_{n \to \infty} \dfrac{f_{n+1,3}}{f_{n,3}} - L_{2^3} \mid < 0.022$. Therefore, one can formulate

Proposition 10.2. Let $f_{n,k} = F_{n \cdot 2^k}$, where $F_{n \cdot 2^k}$ is the n-th term of Fibonacci sieve of order k, and L_n be the n-th Lucas number. Then for any positive number ε there exists number k such that

$$\mid \lim\limits_{n \to \infty} \frac{f_{n+1,k}}{f_{n,k}} - L_{2^k} \mid < \varepsilon.$$

	B	C	D	E	F	G	H
1	1	34	1597	75025	3524578	165580141	7778742049
2							
3	Fibonacci	Lucas	0	*a*		F_{n+1}/F_n	L_{n+1}/L_n
4	1	2	sign slider	47		1	0.5
5	1	1				2	3
6	2	3				1.5	1.33333333
7	3	4				1.66666667	1.75
8	5	7	47			1.6	1.57142857
9	8	11				1.625	1.63636364
10	13	18	value slider			1.61538462	1.61111111
11	21	29	1	*b*		1.61904762	1.62068966
12	34	47	sign slider	-1		1.61764706	1.61702128
13	55	76				1.61818182	1.61842105
14	89	123				1.61797753	1.61788618
15	144	199				1.61805556	1.61809045
16	233	322	1			1.61802575	1.61801242
17	377	521	value slider			1.61803714	1.61804223
18	610	843				1.61803279	1.61803084
19	987	1364				1.61803445	1.61803519
20	1597	2207				1.61803381	1.61803353
21	2584	3571	*k*	2^k		1.61803406	1.61803416
22	4181	5778	3	8		1.61803396	1.61803392
23	6765	9349				1.618034	1.61803401
24	10946	15127				1.61803399	1.61803398

Fig. 10.10. Fibonacci sieve of order three (row 1) and the Golden Ratio approximations (columns G and H).

Remark 10.2. Equations (10.25)-(10.26) represent a special case of Eqs. (10.13)-(10.14) when $a = L_{2^k}$, $b = -1$, $\alpha = F_0$, and $\beta = F_{2^k}$. The spreadsheet that generates solutions to Eqs. (10.25)-(10.26) for different values of k (cell D22) is shown in Fig. 10.10. It should be noted that any number generated by Eqs. (10.25)-(10.26) is a Fibonacci number. Once again, one can use the spreadsheet pictured in Fig. 10.6 to support the last statement. Because, as shown in the spreadsheet of Fig. 10.5, this is not the case for other integer values of a and b in Eq. (10.13), the family of difference equations (10.25)-(10.26) appears to be rather special. The last note, in particular, raises the question whether there exist other values of parameters a and b in Eq. (10.13) that yield *different* Fibonacci numbers only. The use of the word different in the last sentence is due to a trivial sequence 1, 1, 1, 1, ... of the same (Fibonacci) numbers that Eq. (10.13) generates for the initial values $\alpha = \beta = 1$ and for any pair of parameters a and b that are satisfying the relation $a + b = 1$. Indeed, it follows from the last relation that $b = a - 1$ and, therefore, relations (10.13)-(10.14) have the form $f_{n+1} = af_n + (1-a)f_{n-1}, f_0 = f_1 = 1$ implying that $f_n = 1$ for all $n \geq 1$; that is, the sequence 1, 1, 1, 1, ... may be interpreted as a sequence consisting of the same Fibonacci numbers.

	A	B	C	D	E	F	G
1	$f_{n,k}$	1	17167680177565	4.07306E+26	9.7E+39	2.2927E+53	5.43936E+66
2							
3	n	Fibonacci	Lucas	a			$f_{n+1,k}/f_{n,k}$
4	0	1	2	23725150497407			2.618033989
5	1	1	1				6.854101966
6	2	2	3	b			46.97871376
7	3	3	4	-1			2206.999547
8	4	5	7				4870847
9	5	8	11	k	2^k		2.37252E+13
10	6	13	18	6	64		
11	7	21	29				
12	8	34	47				

Fig. 10.11. Modeling $\lim\limits_{n \to \infty} \dfrac{f_{n+1,k}}{f_{n,k}}$ for $1 \leq k \leq 6$.

10.6 Computational Experiments with Fibonacci-Like Sequences

Introducing computational experiment approach into mathematics teacher education, note that, unlike experimentation in mathematical research, the approach, usually, does not offer results that were not possible to obtain in the pre-digital era. For example, as was shown above, through an unsophisticated experimentation with the first two terms of Fibonacci sequence F_n one can come across different number sequences (e.g., Lucas numbers) and discover that, regardless of their values, the ratios F_{n+1}/F_n always tend, as n grows larger and larger, to the Golden Ratio, $\phi = \dfrac{1+\sqrt{5}}{2}$ (Fig. 10.12). As was mentioned in Sec. 4 of this chapter, the ratios of two consecutive terms of any Fibonacci-like sequence tend to the Golden Ratio ϕ as the terms grow large. This computational experiment can be extended to explore the behavior of the fraction $\dfrac{f_{n+2}(x,y)}{f_n(x,y)}$ which is the ratio of two consecutive terms of a subsequence $f_0(x,y), f_2(x,y), f_4(x,y), f_6(x,y),...$ of sequence (10.29) after the elimination of every second term of the sequence.

	A	B	C	D	E	F	G	H
1	1			2			1.5	
2	1	1		1	0.5		3.6	2.4
3	2	2		3	3		5.1	1.41667
4	3	1.5		4	1.33333		8.7	1.70588
5	5	1.66667		7	1.75		13.8	1.58621
6	8	1.6		11	1.57143		22.5	1.63043
7	13	1.625		18	1.63636		36.3	1.61333
8	21	1.61538		29	1.61111		58.8	1.61983
9	34	1.61905		47	1.62069		95.1	1.61735
10	55	1.61765		76	1.61702		153.9	1.6183
11	89	1.61818		123	1.61842		249	1.61793
12	144	1.61798		199	1.61789		402.9	1.61807
13	233	1.61806		322	1.61809		651.9	1.61802
14	377	1.61803		521	1.61801		1054.8	1.61804
15	610	1.61804		843	1.61804		1706.7	1.61803
16	987	1.61803		1364	1.61803		2761.5	1.61803
17	1597	1.61803		2207	1.61804		4468.2	1.61803
18	2584	1.61803		3571	1.61803		7229.7	1.61803
19	4181	1.61803		5778	1.61803		11698	1.61803

Fig. 10.12. The limiting behavior of the sequence F_{n+1}/F_n does not depend on the values of F_0 and F_1.

Using a spreadsheet, one can show experimentally that this new ratio sequence tends to the number 2.618034 (Fig. 10.13) and then prove formally that this number is the approximation to $\dfrac{3+\sqrt{5}}{2} = \lim\limits_{n \to \infty} \dfrac{F_n}{F_{n-2}}$.

Indeed,

$$\lim_{n \to \infty} \frac{f_{n+2}(x,y)}{f_n(x,y)} = \lim_{n \to \infty} \frac{F_n x + F_{n+1} y}{F_{n-2} x + F_{n-1} y} = \lim_{n \to \infty} \left[\frac{F_n}{F_{n-2}} \cdot \frac{x + \dfrac{F_{n+1}}{F_n} y}{x + \dfrac{F_{n-1}}{F_{n-2}} y} \right] = \lim_{n \to \infty} \left[\frac{F_n}{F_{n-2}} \cdot \frac{x + \phi y}{x + \phi y} \right]$$

$$= \lim_{n \to \infty} \frac{F_n}{F_{n-2}} = \lim_{n \to \infty} \frac{F_{n-1} + F_{n-2}}{F_{n-2}} = \lim_{n \to \infty} \left(\frac{F_{n-1}}{F_{n-2}} + 1 \right) = \frac{1+\sqrt{5}}{2} + 1 = \phi + 1.$$

	A	B	C	D	E	F	G	H
1	1			2			1.5	
2	2	2		1	0.5		3.6	2.4
3	5	2.5		1	1		9.3	2.583333
4	13	2.6		2	2		24.3	2.612903
5	34	2.615385		5	2.5		63.6	2.617284
6	89	2.617647		13	2.6		166.5	2.617925
7	233	2.617978		34	2.61538		435.9	2.618018
8	610	2.618026		89	2.61765		1141.2	2.618032
9	1597	2.618033		233	2.61798		2987.7	2.618034
10	4181	2.618034		610	2.61803		7821.9	2.618034
11	10946	2.618034		1597	2.61803		20478	2.618034
12	28657	2.618034		4181	2.61803		53612.1	2.618034
13	75025	2.618034		10946	2.61803		140358.3	2.618034
14	196418	2.618034		28657	2.61803		367462.8	2.618034
15	514229	2.618034		75025	2.61803		962030.1	2.618034
16	1346269	2.618034		196418	2.61803		2518627.5	2.618034
17	3524578	2.618034		514229	2.61803		6593852.4	2.618034
18	9227465	2.618034		1346269	2.61803		17262929.7	2.618034
19	24157817	2.618034		3524578	2.61803		45194936.7	2.618034

Fig. 10.13. The limiting behavior of the sequence F_{n+2} / F_n does not depend on the values of F_0 and F_2.

One can continue this investigation and explore the behavior of the fraction $\dfrac{f_{n+4}(x,y)}{f_n(x,y)}$ — the ratio of two consecutive terms of the subsequence $f_0(x,y), f_4(x,y), f_8(x,y), f_{12}(x,y),...$ of sequence (10.29). This yields the following:

$$\lim_{n\to\infty}\frac{f_{n+4}(x,y)}{f_n(x,y)} = \lim_{n\to\infty}\frac{F_{n+2}x+F_{n+3}y}{F_{n-2}x+F_{n-1}y} = \lim_{n\to\infty}\left[\frac{F_{n+2}}{F_{n-2}}\cdot\frac{x+\dfrac{F_{n+3}}{F_{n+2}}y}{x+\dfrac{F_{n-1}}{F_{n-2}}y}\right] = \lim_{n\to\infty}\left[\frac{F_{n+2}}{F_{n-2}}\cdot\frac{x+\phi y}{x+\phi y}\right]$$

$$=\lim_{n\to\infty}\frac{F_{n+2}}{F_{n-2}} = \lim_{n\to\infty}\frac{F_{n+1}+F_n}{F_{n-2}} = \lim_{n\to\infty}\frac{F_n+F_{n-1}+F_{n-1}+F_{n-2}}{F_{n-2}} = \lim_{n\to\infty}\left[\frac{F_n+F_{n-1}}{F_{n-2}}+\frac{F_{n-1}+F_{n-2}}{F_{n-2}}\right]$$

$$=\lim_{n\to\infty}\left[\frac{F_n}{F_{n-2}}+2\cdot\frac{F_{n-1}}{F_{n-2}}+1\right] = \lim_{n\to\infty}\frac{F_n}{F_{n-2}}+2\phi+1 = \lim_{n\to\infty}\frac{F_{n-1}+F_{n-2}}{F_{n-2}}+2\phi+1$$

$$=\lim_{n\to\infty}\frac{F_{n-1}}{F_{n-2}}+1+2\phi+1 = 3\phi+2.$$

One can use a spreadsheet similar to the one shown in Fig. 10.13 to obtain the number 6.854102 as an approximation to $3\phi+2$. Furthermore,

$$\lim_{n\to\infty}\frac{f_{n+8}(x,y)}{f_n(x,y)} = \lim_{n\to\infty}\frac{F_{n+6}x+F_{n+7}y}{F_{n-2}x+F_{n-1}y} = \lim_{n\to\infty}\left[\frac{F_{n+6}}{F_{n-2}}\cdot\frac{x+\dfrac{F_{n+7}}{F_{n+6}}y}{x+\dfrac{F_{n-1}}{F_{n-2}}y}\right] = \lim_{n\to\infty}\left[\frac{F_{n+6}}{F_{n-2}}\cdot\frac{x+\phi y}{x+\phi y}\right]$$

$$=\lim_{n\to\infty}\frac{F_{n+6}}{F_{n-2}} = \lim_{n\to\infty}\frac{F_{n+5}+F_{n+4}}{F_{n-2}} = \lim_{n\to\infty}\frac{2F_{n+4}+F_{n+3}}{F_{n-2}} = \lim_{n\to\infty}\frac{3F_{n+3}+2F_{n+2}}{F_{n-2}}$$

$$=\lim_{n\to\infty}\frac{3(F_{n+2}+F_{n+1})+2F_{n+2}}{F_{n-2}} = \lim_{n\to\infty}\frac{5F_{n+2}+3F_{n+1}}{F_{n-2}} = \lim_{n\to\infty}\frac{5(F_{n+1}+F_n)+3F_{n+1}}{F_{n-2}}$$

$$=\lim_{n\to\infty}\frac{8F_{n+1}+5F_n}{F_{n-2}} = \lim_{n\to\infty}\frac{8(F_n+F_{n-1})+5F_n}{F_{n-2}} = \lim_{n\to\infty}\frac{13F_n+8F_{n-1}}{F_{n-2}}$$

$$=\lim_{n\to\infty}\frac{13(F_{n-1}+F_{n-2})+8F_{n-1}}{F_{n-2}} = 21\phi+13.$$

Once again, a spreadsheet can be used to arrive to the number 46.978714 as an approximation to $21\phi+13$. The above computations lead to

Proposition 10.3. The following technology-motivated relationship

$$\lim_{n \to \infty} \frac{f_{n+2^k}(x, y)}{f_n(x, y)} = F_{2^k - 1} \cdot \phi + F_{2^k - 2} \tag{10.32}$$

holds true.

Proof. One can use *Maple* (Figs. 10.14 & 10.15) to prove formula (10.32) by the method of mathematical induction (see Sec. 1.7 of Chapter 1). When $k = 1$ in (10.32), the equality

$$\lim_{n \to \infty} \frac{f_{n+2}(x, y)}{f_n(x, y)} = \phi + 1 = F_1 \cdot \phi + F_0,$$

as was demonstrated above, holds true. By defining $P(k) = LI(k) - F(2^k - 1) \cdot GR - F(2^k - 2)$, where $F(n)$ is defined through formula (10.27), GR is the Golden Ratio, $LI(k)$ is defined as the left-hand side of formula (10.32), and $f_n(x, y)$ is defined through formula (10.30), one assumes that formula (10.32) holds true. Then, if one can show that the difference $P(k + 1) - P(k)$ can be simplified to zero, then formula (10.32) holds true when k is replaced by $k + 1$, and this would conclude the inductive transfer in proving formula (10.32). Unfortunately, *Maple* does not simplify the expression $P(k + 1) - P(k)$ to zero itself, but it is not difficult to see that the resulting expression

$$4^{-2^k}(6 + 2\sqrt{5})^{2^k} - 2^{-2^k}(\sqrt{5} + 1)^{2^k} - (\frac{3}{2} + \frac{1}{2}\sqrt{5})^{2^k} + (\frac{1}{2} + \frac{1}{2}\sqrt{5})^{2^k}$$ is equal to

zero (Fig. 10.15). Alternatively, the last expression can be entered into the input box of *Wolfram Alpha* yielding exactly zero (Fig. 10.16). This concludes the proof of Proposition 10.3.

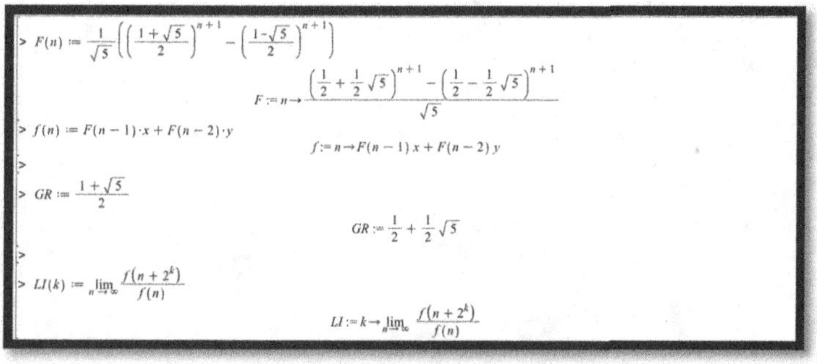

Fig. 10.14. *Maple*-based mathematical induction proof of formula (10.32) — part 1.

$$> P(k) := LI(k) - F(2^k - 1) \cdot GR - F(2^k - 2)$$
$$P := k \to LI(k) - F(2^k - 1)\, GR - F(2^k - 2)$$
$$> P(k+1) - P(k)$$

$$\frac{\left(\left(\sqrt{5}+1\right)^{2^k}\right)^2}{\left(2^{2^k}\right)^2} - \frac{1}{5}\sqrt{5}\left(\left(\frac{1}{2}+\frac{1}{2}\sqrt{5}\right)^{2^k+1} - \left(\frac{1}{2}-\frac{1}{2}\sqrt{5}\right)^{2^k+1}\right)\left(\frac{1}{2}+\frac{1}{2}\sqrt{5}\right) - \frac{1}{5}\sqrt{5}\left(\left(\frac{1}{2}+\frac{1}{2}\sqrt{5}\right)^{2^{k+1}-1}\right.$$

$$\left. - \left(\frac{1}{2}-\frac{1}{2}\sqrt{5}\right)^{2^{k+1}-1}\right) - \frac{\left(\sqrt{5}+1\right)^{2^k}}{2^{2^k}} + \frac{1}{5}\sqrt{5}\left(\left(\frac{1}{2}+\frac{1}{2}\sqrt{5}\right)^{2^k} - \left(\frac{1}{2}-\frac{1}{2}\sqrt{5}\right)^{2^k}\right)\left(\frac{1}{2}+\frac{1}{2}\sqrt{5}\right) + \frac{1}{5}\sqrt{5}\left(\left(\frac{1}{2}\right.\right.$$

$$\left.\left. + \frac{1}{2}\sqrt{5}\right)^{2^k-1} - \left(\frac{1}{2}-\frac{1}{2}\sqrt{5}\right)^{2^k-1}\right)$$

$$> simplify(\%)$$

$$4^{-2^k}\left(6+2\sqrt{5}\right)^{2^k} - 2^{-2^k}\left(\sqrt{5}+1\right)^{2^k} - \left(\frac{3}{2}+\frac{1}{2}\sqrt{5}\right)^{2^k} + \left(\frac{1}{2}+\frac{1}{2}\sqrt{5}\right)^{2^k}$$

Fig. 10.15. *Maple*-based mathematical induction proof of formula (10.32) — part 2.

Input:

$$4^{-2^k}\left(6+2\sqrt{5}\right)^{2^k} - 2^{-2^k}\left(\sqrt{5}+1\right)^{2^k} - \left(\frac{3}{2}+\frac{\sqrt{5}}{2}\right)^{2^k} + \left(\frac{1}{2}+\frac{\sqrt{5}}{2}\right)^{2^k}$$

Result:

0

Fig. 10.16. Complementing *Maple* with *Wolfram Alpha* in showing zero.

Appendix

One Hundred Problems

A.1 Introduction

This appendix includes 100 problems to be solved/explored/ investigated/extended using integrated spreadsheets. Some problems are follow-up tasks on the examples discussed throughout the book. It is expected that when working on those tasks, the reader develops initial understanding through numerical evidence and then, following the approach taken in the book, moves to formal mathematical demonstration supported by symbolic computations as appropriate (e.g., using *Maple*-based mathematical induction proof). Some other problems are purely computational tasks that don't require any generalization to be carried out. Nonetheless, for such tasks generalization may be helpful as a means of constructing a computational environment for solving similar problems. In that way, in the digital era, problem solving can not only motivate but also enhance problem posing. Although many problems included in the appendix can be solved just with pencil and paper, one of the reasons of using technology in mathematical education is to make traditionally difficult problems accessible to as many learners of mathematics as possible. That is why problem solving with integrated spreadsheets is strongly encouraged throughout the appendix.

The problems also differ in length (when a problem consists of several sub-problems) and complexity (either computational or mathematical). Lengthy problems (e.g., 81, 85, 86) can be used for teacher candidates' individual or group projects within a technology-enhanced mathematics education course; more complicated problems (e.g., 82-84, 93-97) can be utilized by practicing teachers for the development of creativity and giftedness of schoolchildren, both inside and outside the classroom. Also, included are problems (e.g., 53, 55, 66) the context of which spans over several chapters of the book; that is why the whole appendix was not divided into specific sections. Finally, problems associated in one or other way with well-known sources are provided with references (e.g., 59, 76, 79).

A.2 Problems

1. Can the number 100 represent the sum of consecutive natural numbers starting from one? Why or why not?

2. Can the number 100 represent the sum of consecutive natural numbers starting from a number other than one? Find all such starting numbers.

3. Can the number 100 represent the sum of any three consecutive natural numbers? Why or why not?

4. Can the number 100 represent the sum of any three consecutive odd/even numbers? Why or why not?

5. Can the number 100 represent the sum of any four consecutive natural numbers? Why or why not?

6. Can the number 100 represent the sum of any four consecutive odd/even numbers? Why or why not?

7. Can the sum of consecutive odd numbers starting from one be a perfect square? Why or why not?

8. Can the sum of consecutive even numbers starting from two be a perfect square? Why or why not?

9. Can the sum of consecutive positive integers starting from one be a power of two? Why or why not?

10. Can the sum of consecutive positive integers starting from a number other than one be a power of two? Why or why not?

11. How many multiples of two among the terms of the arithmetic sequence 1, 8, 15, 22, 29, 36, ... do not exceed 10^3? Answer this question first computationally and then by using algebra.

12. How many multiples of three among the terms of the arithmetic sequence 1, 5, 9, 13, 17, 21, ... do not exceed 10^3? Answer this question first computationally and then by using algebra.

13. How many multiples of five among the terms of the arithmetic sequence 1, 5, 9, 13, 17, 21, 25, ... do not exceed 10^3? Answer this question first computationally and then by using algebra.

14. Using a spreadsheet, explore divisibility of the following polynomials $P(n)$, $n \in N$, by a given number. Then using *Maple*, prove the results of empirical induction by the method of mathematical induction.

(i) $P(n) = n^3 + 11n$ by the number 6;

(ii) $P(n) = n^4 + 6n^3 + 11n^2 + 6n$ by the number 24;

(iii) $P(n) = n^5 - 125n^3 + 4n$ by the number 120;

(iv) $P(n) = n^8 - n^6 - n^4 + n^2$ by the number 1152 when $n = 2k + 1, k \in N$.

15. Using a spreadsheet generate the sequence 1, 2, 3, 4, 5, 6, 7, 1, 2, 3, 4, 5, 6, 7, 1, 2, 3, 4, 5, 6, 7,.... . What are possible concrete situations that could be described by this sequence?

16. Using a spreadsheet generate the sequence 1, 4, 7, 10, 1, 4, 7, 10, 1, 4, 7, 10, ... What are possible concrete situations that could be described by this sequence?

17. Using a spreadsheet generate the sequence 1, 4, 7, 10, 10, 7, 4, 1, 1, 4, 7, 10, 10, 7, 4, 1, 1, 4, 7, 10, ... What are possible concrete situations that could be described by this sequence?

18. Using a spreadsheet, generate the sequence 1, 2, 3, 3, 2, 1, 1, 2, 3, 3, 2, 1, 1, 2, 3, 3, 2, 1,... . What are possible concrete situations that could be described by this sequence?

19. Using a spreadsheet, generate the sequence 5, 5, 6, 5, 1, 7, 5, 5, 6, 5, 3, 1, 5, 5, 6, 5, 1, 7, 5, 5, 6, 5, 3, 1, 5, 5, 6, 5, 1, 7, 5, 5, 6, 5, 3, 1, ... Assign the notes to numbers as follows: "do" to 1, "re" to 2, "mi" to 3, "fa" to 4, "sol" to 5, "la" to 6, and "ti" to 7. Do you hear a familiar piece of music?

20. Find the ratio of the sums of numbers divisible by two in $(n+1) \times (n+1)$ and $n \times n$ addition tables. How does the ratio behave? Construct a graph demonstrating this behavior. Do the same investigation for the multiples of three. Describe any pattern found.

21. Find the ratio of the sums of numbers divisible by two in $(n+1) \times (n+1)$ and $n \times n$ multiplication tables. How does the ratio behave? Construct a graph demonstrating this behavior. Do the same investigation for the multiples of three. Describe any pattern found.

22. Find the size of the smallest checkerboard for which the ratio of the total number of rectangle that it contains to the sum of their semi-perimeters is smaller than 10^{-4}.

23. A number is chosen at random from the $n \times n$ addition table.

    (i) What is the probability that the chosen number is a multiple of two?

    (ii) What is the probability that the chosen number is a multiple of three?

24. A number is chosen at random from the $n \times n$ multiplication table.

(i) What is the probability that the chosen number is a multiple of two?

(ii) What is the probability that the chosen number is a multiple of three?

25. How many products in the $n \times n$ multiplication table are divisible by four?

26. The number of calories in a vanilla ice cream cone depends on the number of grams G of ice cream used. This function C can be expressed using the rule $C(G) = 1.6G + 110$. Using a spreadsheet, construct a computational environment for exploring the situation. Formulate and explore as many questions about the situation as you can and extend the situation in any direction you wish.

27. Melanie won $100,000 in a lottery. She put the money in a savings account that pays 8% interest. At the end of each year she withdraws $15,000 from the account and spends it on a big bash. Using a spreadsheet, determine how many years she can do this before her account balance is down to zero. Investigate the situation by altering the above data. Describe all interesting results that can be found through spreadsheet modeling. Formulate new situations that may be appropriate to explore.

28. Two electric companies measure the amount of electricity its customers use in kilowatt hours (kwh) and charge them according to the following schedules. Company A: the first 12 kwh or less — $4.80; the next 78 kwh — $0.16; the excess above 90 kwh — $0.12 (so that the minimum bill is $4.80). Company B: the first 15 kwh or less — $5.10; the next 70 kwh — $0.13; the excess above 85 kwh — $0.15 (so that the minimum bill is $5.10). Construct a spreadsheet environment for calculating customers' bills for the two companies. Find the amount of electricity for which the bills of these two companies differ by less than five cents. Investigate the situation by altering the schedules. Describe other interesting results that you can find in the spreadsheet environment. Formulate other situations about the charges and bills and explore them.

29. Last Sunday Steve's Video Store rented two types of new movies: *Gone Girl* and *Predestination*. Manager knows that the total sales for renting these movies last Sunday were $100, and that the store rented them in the price range from $3 to $6. Furthermore, it is known that

Gone Girl was a more expensive movie than *Predestination*. Other than that, the manager does not have any record related to the Sunday rent and wants to regain at least part of the renting history by using a spreadsheet. Help the manager find answers to the following questions: What is the maximum number of movies that could have been rented? What are the prices in this case? What is the minimum number of movies (greater than zero) that could have been rented? What are the prices in this case? How many different combinations of $5 *Gone Girl* and $3 *Predestination* movies might there have been rented? What rent prices were not possible within the above range?

30. A machine changed a half-dollar coin into quarters, dimes, and nickels. Assuming that there is equally likely to get any combination of the coins, find the probability that there were no nickels in the change.

31. A machine changed a dollar coin into half-dollars, quarters, and dimes. Assuming that there is equally likely to get any combination of the coins, find the probability that there were no dimes in the change.

32. How many ways can one change a $100 bill using $20, $10, and $5 bills? Assuming that it is equally likely to receive change in any combination of the bills, what is the probability of receiving exactly six banknotes?

33. How many ways can one make a quarter out of pennies, nickels, and dimes? If a machine randomly changes quarters into pennies, nickels, and dimes, what is the probability that

  (i) there are no pennies in a change?
  (ii) there are no nickels in a change?
  (iii) there are no dimes in a change?

34. Manager of a grocery store is looking for someone who can design a machine which can make change for customers. The machine should display how many bills or coins of each denomination to return, and should return the fewest possible pieces of money. Design a spreadsheet-based program that can operate such a machine with bills and coins of denomination $10, $5, $1, 25 cents, 10 cents, 5 cents, and 1 cent.

35. A census taker asked the farmer the ages of his three daughters. The farmer told him that the product of their ages is 72 and the sum of their ages is a square number. The census taker performed some computations and told the farmer that he could not tell the ages of the daughters. The farmer said, "I forgot to tell you that the youngest likes petting dogs".

This helped the census taker and he now knew the ages of the three daughters. What are the ages of the three daughters?

36. A census taker asked the farmer the ages of his three daughters. The farmer told him that the product of their ages is 80 and the sum of their ages is a square number. The census taker performed some computations and told the farmer that he could not tell the ages of the daughters. The farmer said, "I forgot to tell you that the oldest attends elementary school". This helped the census taker and he now knew the ages of the three daughters. What are the ages of the three daughters?

37. A census taker asked the farmer the ages of his three daughters. The farmer told him that the product of their ages is 36 and the sum of their ages is the house number. The census taker performed some computations and then looked at the house number. At this point, he told the farmer that he still could not tell the ages of the daughters. The farmer said, "I forgot to tell you that the youngest likes chocolate milk". This helped the census taker and he now knew the ages of the three daughters. What are the ages of the three daughters?

38. The age of a man is the same as his wife's age with the digits reversed. The sum of their ages is 99 and the man is nine years older than his wife. How old is the man? Develop a spreadsheet environment enabling posing new problems of that type. Can the sum of their ages be any number? Why or why not? By altering the sum of their ages, find all possible solutions provided that the man is nine years older than his wife. Is it possible for a man to be eight years older than his wife provided that their ages are represented by numbers with the digits swapped? Under which conditions a man may be eight years older than his wife?

39. The sum of ages of three children is 32. The age of the oldest is twice the age of the youngest. The ages of the two older children differ by three years. What is the age of the youngest child? Develop a spreadsheet environment for posing new problems of that kind.

40. It takes 43 cents in postage to mail a letter. A post office has stamps of denomination five cents, seven cents, and ten cents. How many combinations of the stamps could Anna buy to send a letter?

41. At the post office, a person spent a total of $2.00 to get some 29¢-stamps and some 5¢-stamps, and received no change. How many 5¢-stamps did the person buy?

42. Find all ways to spend $25 at a book sale where books are priced $4 and $5.

43. How many ways can one spend at most $30 by buying books priced $5 and $6?

44. A post office has a large number of stamps in two different denominations. Using combinations of only these two stamps, one cannot pay twenty-seven cents postage but can pay any amount greater than twenty-seven cents. What are the two denominations of the stamps?

45. One afternoon Bobby decided he would count the traffic going over the Long Bridge. In the evening, Bobby's father asked how many bicycles, tricycles, and cars he counted. Bobby remembered that he saw not fewer than 12 and not greater than 14 vehicles, not fewer than 45 and not greater than 50 wheels, and that it was the same number of bicycles as tricycles. This information was enough for Bobby's father to get the exact count of each of the three vehicles. Use a spreadsheet to process this information and yield an answer. Then verify your answer by using *Wolfram Alpha*. Use a spreadsheet to pose new problems of that kind. Solve your own problems.

46. How many ways can Julie spend $30 to buy books priced $10, $9, $8, and $7?

47. From a pile of 100 pennies (P), 100 nickels (N), and 100 dimes (D), select 21 coins which have a total value of exactly $1.00. In your selection you must also use at least one coin of each type. How many coins of each of the three types (P, N, D) should be selected?

48. Consider the equation $ax + by + cz = n$ with whole number coefficients $a, b, c, n; n > 0$. Let $D(n; a, b, c)$ be a function representing the number of ways the number n can be partitioned into the summands $a, b,$ and c (referred to as denumerant in Chapter 3).

   (i)   Is there a triple (a, b, c) for which the function assumes values of consecutive natural numbers with each value taken only once?

   (ii) Is there a triple (a, b, c) for which the function assumes values coming in pairs of natural numbers?

   (iii) Is there a triple (a, b, c) for which the function assumes values of an arithmetic sequence?

   (iv) Is there a triple (a, b, c) for which the function assumes values of a geometric sequence?

(v) Which triple (a, b, c) gives the longest string in terms of the sequence $x_n = D(n;a,b,c)$ being in arithmetic progression?

(vi) Which triple (a, b, c) gives the longest string in terms of the sequence $x_n = D(n;a,b,c)$ being in geometric progression?

(vii) Is there a triple (a, b, c) for which the function $D(n;a,b,c)$ assumes values coming in pairs of equal numbers?

49. Three segments are given whose lengths are 2, 3, and 5 centimeters. Using any of the given lengths as many times as you wish, determine how many segments of length 20 centimeters can be constructed.

50. Three segments are given whose lengths are 3, 4, and 5 centimeters. Using any of the given lengths as many times as you wish, determine how many segments of length not greater than 16 centimeters can be constructed.

51. Four segments are given whose lengths are 1, 2, 3, and 4 centimeters. Using any of the given lengths as many times as you wish, determine how many segments of length 10 centimeters can be constructed.

52. Find all four-digit numbers divisible by 45 and having 9 and 7 as the second and the third digit, respectively.

53. Find all four-digit prime numbers having 9 and 7 as the second and the third digit, respectively.

54. The number 149 and its digit-reversed associate 941 are both prime numbers. Find all three-digit numbers with such a property. Are there four-digit numbers that possess the same property? Why or why not? Design an environment that displays pairs of such digit-reversed prime numbers.

55. A combination lock of Jimmy's new briefcase has three digital disks, each containing all the digits from 0 through 9. Jimmy forgot his secrete code; yet he remembers that the code is a three-digit prime number palindrome. Jimmy knows that the lock is designed in such a way that it allows one to guess a code only a limited number of times — it locks up (breaks) forever after the number of trials becomes greater than the total number of three-digit prime number palindromes. Under the above conditions, investigate the following queries.

(i) After how many trials would the combination lock break?

(ii) What is the code for this lock, if in addition to the above information, Jimmy recalls that the sum of digits of his code is a square number?

(iii) Do you need to explore three-digit palindromes in the range 500-599 using a spreadsheet? Why or why not?
(iv) What is the smallest number of primes that may be used to investigate this situation using a spreadsheet?
(v) Formulate new questions about this situation (or suggest its extension/modification) and answer your own questions[s].

56. Use a spreadsheet to find all ways to represent the number 221 as a sum of two squares. Represent 221 as a product of two prime factors using Euler's factorization method (Sec. 4.7 of Chapter 4). What do the factors have in common?

57. Use a spreadsheet to find all ways to represent the number 1105 as a sum of two squares. Represent 1105 as a product of two factors using Euler's factorization method (Sec. 4.7 of Chapter 4). How many ways can this be done? Find the prime factorization of the number 1105. What do the factors have in common?

58. Using a spreadsheet (see Chapter 4, Fig. 4.6), test that the following eight-digit number 73,939,133 is a prime. This number is special because all the numbers 7, 73, 739, 7393, 73939, 739391, 7393913 that gradually build up to it from the left to the right are primes as well. A similar property is possessed by a seven-digit number 3,333,331, because all the numbers 31, 331, 3331, 33331, 333331 that gradually build up to it from the right to the left are primes as well. Using an integrated spreadsheet, find out whether there exists another such number with not fewer than four digits or prove that such number does not exist. Extend the number

73,939,133 by adding one or more digits to the right to see whether there exists a prime number extension of this eight-digit number. Carry out the same exploration for the number 3,333,3331.

59. Following recess the 1000 students of a school lined up and entered the school as follows: The first student opened all of the 1000 lockers in the school. The second student closed all lockers with even numbers. The third student "changed" all lockers that were numbered with multiples of three by closing those that were open and opening those that were closed. The fourth student changed each locker numbered with a multiple of four, and so on. Alter all 1000 students had entered the building in this fashion, which lockers were left open? (See [Musser and Burger, 1997, p. 201]).

60. How many subgrids of a 73-cell grid constitute its whole number percentage, provided that shape and location of a subgrid is immaterial?

61. How many subgrids of a 84-cell grid constitute its whole number percentage, provided that shape and location of a subgrid is immaterial?

62. How many subgrids of a 76-cell grid constitute its whole number percentage of a larger, perhaps, a "hypothetical" grid for which the 76-cell grid is a 19% part, provided that shape and location of a subgrid is immaterial?

63. How many subgrids of a 64-cell grid constitute its whole number percentage of a larger, perhaps, a "hypothetical" grid for which the 64-cell grid is a 28% part, provided that shape and location of a subgrid is immaterial?

64. Find an integer-sided rectangle with the smallest perimeter which area plus the excess of the length over the width is equal to 281. Is such a rectangle unique? If so, what is special about this rectangle?

65. In Chapter 6 (Proposition 6.1), the general solution to (Pythagorean) Eq. (6.1) was found in the form of Eqs. (6.5).

(i) For which values of the generators m and n does the triple $(m^2 - n^2, 2mn, m^2 + n^2)$ become the triple $(m + n, 2mn, m^2 + n^2)$?

(ii) How are the generators m and n related when the difference between two elements of a Pythagorean triple is equal to one?

(iii) How are the generators m and n related when the difference between two elements of a Pythagorean triple is equal to two?

(iv) Is it possible to get a primitive Pythagorean triple if the generators m and n have a common factor? Why or why not?

(v) Using an integrated spreadsheet, locate a Pythagorean triple with 43 as the largest element. Do the same for 67, 71, 79, 83, 103. What do all these numbers have in common? In particular, they can be characterized as odd numbers. What distinguishes them from all other odd numbers?

(vi) Is it true that for every primitive Pythagorean triple, its largest element is either a prime number of the form $4n + 1$ or a multiple of such primes?

66. Consider the spreadsheet shown in Fig. 6.10 of Chapter 6 that generates 90°-triples through formulas (6.5). Assuming $2 \le m \le 20$ and using Proposition 9.4 of Chapter 9, answer the following questions:

(i) How many appearances of triangular numbers not greater than 325 are there among the triples?

(ii) How many appearances of square numbers not greater than 289 are there among the triples?

(iii) How many appearances of pentagonal numbers not greater than 425 are there among the triples?

67. Two two-digit numbers are written down one after another so that they constitute a four-digit number which is divisible by their product. Find these numbers.

68. From a three-digit number subtract the sum of its digits to get a new number. Then do the same with this new number. Is it possible, by continuing this process, to reach zero? Construct an environment allowing one to determine the number of steps required to reach zero through this process. How does the number of steps depend on the starting number? Construct the graph of the function that relates a three-digit number to the number of steps required to reach zero through this process.

69. The sum of digits of all page numbers in a book equals 3684. Find the number of pages in this book. Is it possible to find a book with such a sum equals 200? Why or why not? Construct a graph of the function that relates the sum of digits of all page numbers in a book to the number of pages in this book. Consider the range [100, 1000].

70. The sum of the digits of all page numbers in a book is nine times as much as the number of pages. Find the number of pages in such a book. Replace nine by eleven and answer the same question. Replace nine by an arbitrary number *n* and explore for what values of *n* the problem has a solution.

71. A book has 500 pages numbered 1, 2, 3, … and so on. How many times does the digit 1 appear in the page numbers? Explore other digits in terms of the frequency of their appearances as part of a page number.

72. When opening a mathematics textbook, Sean noted that the product of the sums of digits of two consecutive page numbers facing him is equal to 30. How many such pairs of pages can be found in this book? Replace 30 by another number, say, 20. Is there a solution to this problem with the number 20? Develop a function that relates the product of the sums of digits of two page numbers to the number of solutions that such a problem has.

73. The serial number of a camera is a four-digit number smaller than 5000 and it contains the digits 2, 3, 5, and 8. The "3" is next to the "8", the "2" is not next to the "3", and the "5" is not next to the "2". Use a spreadsheet to find the serial number.

74. The sum of digits of a three-digit number 104 is 5. How many different three-digit numbers, including 104, has the number 5 as the sum of digits? Develop a spreadsheet environment enabling one to graph the dependence of the sum of digits on the number of numbers with this sum of digits.

75. A man received a check for a certain amount of money, but on cashing it the teller mistook the number of dollars for the number of cents and conversely. Not noticing this, the man spent 68 cents and discovered to his surprise that he had twice as much money as the check was originally drawn for. Use a spreadsheet to determine the amount of money for which the check could have been written.

76. The existence of integers for which the sum of their cubed digits equals the integer itself was probably first mentioned by Hoppenot [1937]. For example, 153 is such number because $1^3 + 5^3 + 3^3 = 153$. These numbers are called narcissistic/Armstrong numbers, perfect digital variants, or cube attractors. Becker [1984] called the problem of exploring integers with respect to this property the cubed digits problem report that and it was used for many years in the work with in-service teachers [Carmony *et al.*, 1984]. Find all such cube attractors. Try to find similar patterns associated with the exponents 4 and 5.

77. What is the smallest two-digit prime number for which the product of its digits is equal to the sum of its digits plus seven? Formulate a new problem of that type and solve it using a spreadsheet.

78. The cube of a two-digit number has five digits all of which are not only different but don't include the digits of the original (two-digit) number. Find the five-digit number.

79. The number 1729 is the smallest number represented as a sum of two cubes in two different ways. Find the two representations. (See [Silverman, 1993]).

80. Using the digits 1, 2, 3, 4, 5, 6, find all two-digit numbers with the difference two between the second and the first digit.

81. Hao is a student in a teacher education program. He has a pre-student teaching field experience at a local school. Last week, Hao was preparing a mathematics lesson trying to find a grade-appropriate problem for his students. Eventually he came across the following problem: *The sum of digits of all page numbers in Sara's new book equals 1062. How many pages are in the book?* Hao believed that this problem was too difficult for the students and slightly modified it by making the sum a two-digit number: *The sum of digits of all page numbers in Sara's new book equals 62. How many pages are in the book?*

Hao suggested to the students the following problem-solving strategy: use a real book and create a two-column chart so that in the first column they list page numbers starting from the first page and in the second column they list the sums of digits of related page numbers. All of a sudden one of the students raised her hand and said that such a book does not exist unless someone has torn out a page from it. Whispered Hao: "How come? These kids really drive me crazy. I was always picky about non-routine problems. Surprise, making the sum smaller did not simplify the problem." Puzzled by the student's response, Hao looked at the host teacher, hoping to get help. But this was a mini-lesson and it was over. "We will continue working on this problem tomorrow," he said to the students. Help Hao to be better prepared for tomorrow's class by answering the following questions:

    (i) What is wrong with the book from Hao's modified problem?

    (ii) What did the student mean by a torn out (missing) page?

    (iii) Find the page numbers for this missing page. Is such a page unique? Why or why not?

(iv) Solve the original problem about Sara's book.

(v) Create a similar problem and solve it. (As Hao's field experience in mathematics indicates, problem posing and problem solving are two sides of the same coin.)

82.[1] Three men and a monkey gather bananas all day and then fall asleep. During the night each man wakes up in turn and, after giving one banana to the monkey, he removes and hides one third of the pile of bananas for himself. Assuming that each man wakes up only once during the night, find a number of bananas originally gathered so that all divisions came out in integers and give a minimum number of remaining bananas.

83. Four men and a monkey gather bananas all day and then fall asleep. During the night each man wakes up in turn and, after giving one banana to the monkey, he removes and hides one fourth of the pile of bananas for himself. Assuming that each man wakes up only once during the night, find a number of bananas originally gathered so that all divisions came out in integers and give a minimum number of remaining bananas.

84. Four men and a monkey gather bananas all day and then fall asleep. During the night each man wakes up in turn and, after giving two bananas to the monkey, he removes and hides one fourth of the pile of bananas for himself. Assuming that each man wakes up only once during the night, find a number of bananas originally gathered so that all divisions came out in integers and give a minimum number of remaining bananas.

85. Once upon a time, two students met at the library on Monday. They wanted to study a geometry book together; however, there was only a single copy of the book in the library. A librarian said she would need to know specific days of the study in order to put the book on reserve for the students on those days. She noted also, that the library is closed one day a week: one week on Sunday and another week on Monday (the Sunday-Monday-Sunday-Monday pattern of the closure days). The first student said to the librarian: "I will keep coming here every second day." The second student said to the librarian: "And I will keep coming here every third day." They both added: "If our visit falls on Sunday, we will come on the next day, so please start scheduling our visits over from that day." Create a schedule for the librarian to put the book on reserve for

---

[1] For problems 82-84 see Gardner [1961], Pask [1998], Abramovich and Cho [2015].

the students. How many days during the 90 days sequence would such a schedule allow the students to get the book from reserve?

86. Three students met at the swimming pool. This is the best place on our campus, said one of them. I will keep coming here every second day. The second student said that he does not have that much time for sport but still, he will keep coming to the pool every third day. The third student said that because of his part-time job he would be able to attend the pool every fourth day only. Manager of the facility overheard their conversation and noted that the pool is closed on Wednesday. The students decided that if a visit falls on Wednesday, they will come on the next day and start scheduling visits over from that day. Once on Thursday the three students met at the swimming again. Which day a week did the above conversation happen?

Explore the following variations of the swimming pool problem:

(i) What if the pool is closed on Thursday? How would this change affect the days when the three students can meet together?

(ii) Explore the situation when one week the pool is closed on Wednesday, and on the next week it is closed on Thursday (these days can be varied).

(iii) Explore the situation when the pool is closed every second (third) Wednesday (Thursday).

87. Show that there are pairs of different triangular numbers whose product is a square number. For example, $t_2 t_{24} = 30^2$. Use a spreadsheet to discover this property.

88. Using a spreadsheet, generate consecutive triangular numbers starting from one. Divide each triangular number by 11 and display remainders only. Do you see any pattern among the remainders? Describe the pattern found. Try other divisors.

89. Using a spreadsheet, generate consecutive pentagonal numbers starting from one. Divide each pentagonal number by 11 and display remainders only. Do you see any pattern among the remainders? Describe the pattern found. Try other divisors.

90. This situation often arises when computers are used to play games. One player enters a number between, say, 1 and 50 without letting a second player see it, and then the second player tries to guess the number by entering a number into computer. Computer counts the number of guesses and if the player guesses the number in no more than 10 tries, it displays the message "You win." However, after 10 unsuccessful guesses computer displays the message "You lose." Write a spreadsheet program and construct an environment for such a game.

91. Jack has a new toy called *reach-a-number machine* which requires the use of nickels and dimes. The machine displays number 1 for free and then it adds 4 for a nickel and multiply by 3 for a dime. Jack wants to reach the number 101. How should he play in order to minimize costs? In other words, how should Jack play in order to reach 101 for the least amount of money if he is allowed to start from 1, add 4, and multiply by 3 only?

92. A college bookstore has textbook sale. As part of the sale, manager announced that any textbook book a student could get for free if he or she can outsmart the store's famous "Free Book" machine. The machine displays a book's price and offers a student to insert any number of dollars in the range $1 through $5 into the money slot. The machine then adds some amount of dollars in the same range ($1 through $5) and tells a student the resulting sum. If this sum is less than the price of a book, the student-machine interaction continues in the same vein. If a student reaches the book's price first, the machine refunds student's money and drops a free textbook into the basket. If the machine reaches the price first, a student pays whole price for a textbook. Suppose you need a $20 textbook. Develop a strategy which would allow you to get the textbook for free. Describe your strategy. If your strategy turns out to be successful, can you help a friend to get $22 textbook for free? What about helping a friend who wants to get $24 textbook for free?

93. Using a spreadsheet first verify and then prove that the sequence of numbers 210, 20100, 2001000, ..., $2\underbrace{00...0}_{n}1\underbrace{00...0}_{n+1}$, ... consist of triangular numbers only. What is the rank of the number with $2n + 1$ zeros?

94. Using a spreadsheet first verify and then prove that the sequence of numbers 820, 80200, 8002000, ..., $8\underbrace{00...0}_{n}2\underbrace{00...0}_{n+1}$, ... consist of triangular numbers only. What is the rank of the number with $2n + 1$ zeros?

95. Using a spreadsheet first verify and then prove that the sequence of numbers 6, 561, 55611, ..., $\underbrace{55...56}_{n}\underbrace{11...1}_{n}$, ... consist of triangular numbers only. What is the rank of the number with n fives and n ones?

96. Using a spreadsheet first verify and then prove that the sequence of numbers 45, 4950, 499500, ..., $4\underbrace{99...95}_{n}\underbrace{00...0}_{n}$, ... consist of triangular numbers only. What is the rank of the number with n nines and n zeros?

97. Using a spreadsheet first verify and then prove that the sequence of numbers 49, 4489, 444889, ..., $\underbrace{44...4}_{n}\underbrace{88...8}_{n-1}9$, ... consists of square numbers only. What is the rank of the number with n fours and $n-1$ eights?

98. Using an integrated spreadsheet find several hexagonal numbers that are also square numbers.

99. Using an integrated spreadsheet find several pentagonal numbers that are also square numbers.

100. In Chapter 5 the function $T(p) = GCD(1, p) + GCD(2, p) + ... + GCD(p, p)$ was defined. Resolve the following queries.

(i) Does there exist p such that $T(p) = T(p+1)$? Why or why not?

(ii) In the range $1 < p < 100$ find all solutions to the equation $T(p) = T(p+2)$.

(iii) In the range $1 < p < 100$ find all solutions to the equation $T(p+1) - T(p) = 1$.

(iv) In the range $1 < p < 100$ find all solutions to the equation $|T(p+1) - T(p)| = p$.

(v) In the range $1 < p < 100$ find all solutions to the equation $T(p+2) - T(p) = 2$.

(vi) In the range $1 < p < 100$ find all solutions to the equation $|T(p+2) - T(p)| = 2$.

(vii) Is it possible to locate a triple of numbers (p_1, p_2, p_3) for which $T(p_1) = T(p_2) = T(p_3)$?

Bibliography

Abramovich, S. (1999). Revisiting an ancient problem through contemporary discourse, *School Science and Mathematics*, 99(3), pp. 148-155.

Abramovich, S. (2000). Mathematical concepts as emerging tools in computing applications, *Journal of Computers in Mathematics and Science Teaching*, 19(1), pp. 21-46.

Abramovich, S. (2005). Inequalities and spreadsheet modeling, *Spreadsheets in Education*, 2(1), Article 1. Available at: http://epublications.bond.edu.au/ejsie/vol2/iss1/1.

Abramovich, S. (2006). Spreadsheet modeling as a didactical framework for inequality-based reduction, *International Journal of Mathematical Education in Science and Technology*, 37(5), pp. 527-541.

Abramovich, S. (2009). Hidden mathematics curriculum of teacher education: an example, *PRIMUS — Problems, Resources, and Issues in Mathematics Undergraduate Studies*, 19(1), pp. 39-56.

Abramovich, S. (2011). *Computer-Enabled Mathematics: Integrating Experiment and Theory in Teacher Education*, (Nova Science Publishers, Hauppauge, NY).

Abramovich, S. (2014). Revisiting mathematical problem solving and posing in the digital era: toward pedagogically sound uses of modern technology, *International Journal of Mathematical Education in Science and Technology*, 45(7), pp. 1034-1052.

Abramovich, S. and Brantlinger, A. (2004). Technology-motivated teaching of topics in number theory through a tool kit approach, *International Journal of Mathematical Education in Science and Technology*, 35(3), pp. 317-333.

Abramovich, S. and Cho, E. K. (2008). On mathematical problem posing by elementary pre-teachers: the case of spreadsheets, *Spreadsheets in Education*, 3(1), Article 1. Available at: http://epublications.bond.edu.au/ejsie/vol3/iss1/1.

Abramovich, S. and Cho, E. K. (2015). Using digital technology for mathematical problem posing. In J. Cai, N. Ellerton, and F. M. Singer (Eds), *Mathematical Problem Posing: From Research to Effective Practice*, (Springer, New York) pp. 71-102.

Abramovich, S., Easton, J. and Hayes, V. O. (2014). Integrated spreadsheets as learning environments for young children, *Spreadsheets in Education*, 7(2), Article 3. Available at: http://epublications.bond.edu.au/ejsie/vol7/iss2/3.

Abramovich, S., Fujii, T. and Wilson, J. (1995). Multiple-application medium for the study of polygonal numbers, *Journal of Computers in Mathematics and Science Teaching*, 14(4), pp. 521-557.

Abramovich, S. and Leonov, G. A. (2009). Spreadsheets and the discovery of new knowledge, *Spreadsheets in Education*, 3(2), Article 1. Available at: http://epublications.bond.edu.au/ejsie/vol3/iss2/1/.

Abramovich, S. and Lyandres, V. (2009). Encouraging interest of K-12 students and teachers in the STEM disciplines: an example, *International Journal of Experimental Education*, 2, pp. 13-20.

Abramovich, S. and Sugden, S. (2008). Revisiting Polya's summation techniques using a spreadsheet: from addition tables to Bernoulli polynomials, *Spreadsheets in Education*, 2(3), Article 4. Available at: http://epublications.bond.edu.au/ejsie/vol2/iss3/4.

Advisory Committee on Mathematics Education. (2011). *Mathematical Needs: The Mathematical Needs of Learners*, (The Royal Society, London). Available at: http://www.nuffieldfoundation.org/sites/default/files/files/ACME_Theme_B_final.pdf.

Alfors, L. (1962). On the mathematics curriculum of the high school, *The American Mathematical Monthly*, 69(3), pp. 189-193.

Baker J. E. (2007). Excel and the Goldbach comet, *Spreadsheets in Education*, 2(2), Article 2. Available at: http://epublications.bond.edu.au/ejsie/vol2/iss2/2.

Baker J. E. (2013). Pascal pyramids: a mathematical exploration using spreadsheets, *Spreadsheets in Education*, 6(2), Article 3. Available at: http://epublications.bond.edu.au/ejsie/vol6/iss2/2.

Ball, D. L. (2000). Bridging practices: intertwining content and pedagogy in teaching and learning to teach, *Journal of Teacher Education*, 51(3), pp. 241-247.

Beberman, M. (1964). Statement of the problem. In D. Friedman (Ed.), *The Role of Applications in a Secondary School Mathematics*

Curriculum (pp. 1-13), Proceedings of a UICSM conference, Monticello, Ill., 14-19 February 1963. Urbana, Ill.: UICSM.

Becker, J. P. (1984). Integrating problem solving into mathematics teaching. In T. Kawaguchi (Ed.), *Proceedings of ICMI — JSME Regional Conference on Mathematical Education*, (Japan Society of Mathematics Education, Tokyo) pp. 79-94.

Beiler, A. H. (1964). *Recreations in the Theory of Numbers: The Queen of Mathematics Entertains*, (Dover, New York).

Ben-Chaim, D., Lappan, G. and Houang, R. T. (1989). The role of visualization in the middle school mathematics curriculum, *Focus on Learning Problems in Mathematics*, 11(1), pp. 49-60.

Beth, E. W. (1966). Strict demonstration and heuristic procedures, In E. W. Beth and J. Piaget, *Mathematical Epistemology and Psychology*, (Reidel, Dordrecht, The Netherlands) pp. 86-100.

Blum, W. (2002). ICMI Study 14: applications and modeling in mathematics education — Discussion document, *Educational Studies in Mathematics*, 51(1-2), pp. 149-171.

BrainPop UK. (2014). Fibonacci sequence [on-line materials], (Author, Oxford, UK). Available at: http://www.brainpop.co.uk/maths/numberandcalculation/fibonaccisequence/preview.weml.

Brown, S. I. and Walter, M. I. (1990). *The Art of Problem Posing*, (Lawrence Erlbaum Associates, Hillsdale, NJ).

Buddenhagen, J., Ford, C. and May, M. (1992). Nice cubic polynomials, Pythagorean triples, and the Law of Cosines, *Mathematics Magazine*, 64(4), pp. 244-249.

Burkert, W. (1972). *Lore and Science in Ancient Pythagoreanism*, (Harvard University Press, Cambridge, MA).

Cai, J., Ellerton, N. and Singer F. M. (Eds). (2015). *Mathematical Problem Posing: From Research to Effective Practice*, (Springer, New York).

Calder, N. (2010). Affordances of spreadsheets in mathematical investigation: potentialities for learning, *Spreadsheets in Education*, 3(3), Article 4. Available at: http://epublications.bond.edu.au/ejsie/vol3/iss3/4.

Carmony, L., McGlinn, R., Becker, J. P. and Millman, A. (1984). *Problem Solving in Apple Pascal*, (Computer Science Press, Rockland, MD).

Char, B. W., Geddes, K. O., Gonnet, G. H., Leong, B. L., Monagan, M. B. and Watt, S. M. (1991). *Maple V Language Reference Manual*, (Springer, New York).

Cockcroft Report. (1982). *Mathematics Counts. Report of the Committee of Inquiry into the Teaching of Mathematics in Schools Under the Chairmanship of Dr WH Cockcroft*, (Her Majesty's Stationary Office, London).

Cole, M. and Griffin, P. (1980). Cultural amplifiers reconsidered. In D. R. Olson (Ed.), *The Social Foundations of Language and Thought*, (Norton, New York) pp. 343-363.

Common Core State Standards. (2010). *Common Core Standards Initiative: Preparing America's Students for College and Career* [on-line materials]. Available at: http://www.corestandards.org.

Comtet, L. (1974). *Advanced Combinatorics*, (Reidel, Dordrecht, The Netherlands).

Conference Board of the Mathematical Sciences. (2001). *The Mathematical Education of Teachers*, (The Mathematical Association of America, Washington, D.C.).

Conference Board of the Mathematical Sciences. (2012). *The Mathematical Education of Teachers II*, (The Mathematical Association of America, Washington, D.C.).

Conway, J. H. and Guy, R. K. (1996). *The Book of Numbers*, (Copernicus, New York).

Cooke, R. (2010). Life on the mathematical frontier: legendary figures and their adventures, *Notices of the American Mathematical Society*, 57(4), pp. 464-475.

Cuoco, A. (2001). Mathematics for teaching, *Notices of the American Mathematical Society*, 48(2), pp. 168-174.

Department for Education. (2013a). Mathematics: Programme of Study for Key Stage 4, Crown copyright. Available at: http://www.education.gov.uk/nationalcurriculum2014.

Department for Education. (2013b). National Curriculum in England: Mathematics Programmes of Study, Crown copyright. Available at:

https://www.gov.uk/government/publications/national-curriculum-
in-england-mathematics-programmes-of-study.

Department for Education (2014). *National Curriculum in England:
Mathematics Programmes of Study* (updated 16 July 2014), Crown
copyright. Available at:
https://www.gov.uk/government/publications/national-curriculum-
in-england-mathematics-programmes-of-study.

Dickson, L. E. (1971). *History of the Theory of Numbers, vol. II,*
(Chelsea Publishing Company, New York).

Doerr, H. M. and English, L. D. (2003). A modeling perspective on
students' mathematical reasoning about data, *Journal for Research in
Mathematics Education,* 34(2), pp. 110-136.

Engel, A. (1968). Systematic use of applications in mathematics training,
Educational Studies in Mathematics, 1(1-2), pp. 202-221.

Euler, L. (1849). Regula facilis problemata Diophantea per numeros
integros expedite resolvendi (An easy rule for Diophantine problems
to be resolved expeditiously in integers). In *Leonhardi Euleri
Commentationes Arithmeticae* (in Latin, composed by P. H. Fuss
and N. Fuss — great-grandchildren of Euler), (The Imperial St.
Petersburg Academy of Sciences, St. Petersburg, Russia) pp. 263-
269.

Expert Panel on Student Success in Ontario. (2004). *Leading Math
Success: Mathematical Literacy, Grades 7-12,* (Ontario Ministry of
Education, Toronto).

Freudenthal, H. (1978). *Weeding and Sowing,* (Kluwer, Dordrecht, The
Netherlands).

Freudenthal, H. (1983). *Didactical Phenomenology of Mathematical
Structures,* (Reidel, Dordrecht, The Netherlands).

Frykholm, J., Vierling, L. and Glasson, G. (2005). Lessons learned:
Integrating elementary mathematics and science. In D. F. Berlin and
A. L. White (Eds), *Collaboration for the Global Improvement of
Science and Mathematics Education,* International Consortium for
Research in Science and Mathematics Education, Columbus, OH)
pp. 1-19.

Gardner, M. (1961). *More Mathematical Puzzles and Diversions,*
(Penguin Books, New York).

Gauss, C. F. (1966). *Disquisitiones Arithmeticae* (translated from Latin by A. A. Clarke), (Yale University Press, New Haven, CT).

Getzels, J. W. and Jackson, P. W. (1962). *Creativity and Intelligence: Exploration with Gifted Students*, (John Wiley & Sons, New York).

Gruenberger, F. (1984). How to handle numbers with thousands of digits, and why one might want to, *Scientific American*, 250(4), pp. 19-26.

Guin, D. and Trouche, L. (1999). The complex process of converting tools into mathematical instruments: the case of calculators, *International Journal of Computers for Mathematical Learning*, 3(3), pp. 195-227.

Harrison, J. (2008). Formal proof — theory and practice, *Notices of the American Mathematical Society*, 55(11), pp. 1395-1406.

Haspekian, M. (2005). An "instrumental approach" to study the integration of a computer tool into mathematics teaching: the case of spreadsheets, *International Journal of Computers for Mathematical Learning*, 10(2), pp. 109-141.

Hitt, F. (1994). Visualization, anchorage, availability and natural image: polygonal numbers in computer environments, *International Journal of Mathematical Education in Science and Technology*, 25(3), pp. 447-55.

Hoppenot, F. (1937). Courrier du "Sphinx". Sphinx 5, Bruxelles, p. 72.

Hoyles, C. (1994). Computer-based microworlds: A radical vision or a Trojan mouse? In D. F. Robitaille, D. H. Wheeler, and C. Kieran (Eds), *Selected Lectures from the 7th International Congress on Mathematical Education*, (Les Presses de l'Université Laval) pp. 171-182.

Kaprekar, D. (1949). Another solitaire game, *Scripta Mathematica*, 15, pp. 244-245.

Kline, M. (1972). *Mathematical Thought from Ancient to Modern Times*, (Oxford University Press, New York).

Kline, M. (1985). *Mathematics for the Non-mathematician*, (Dover, New York).

Knuth, E. J. (2002). Secondary school mathematics teachers' conceptions of proof, *Journal for Research in Mathematics Education*, 33(5), pp. 379-405.

Korovkin, P. P. (1961). *Inequalities*, (Blaisdell Publishing Company, New York).

Koshy, T. (2001). *Fibonacci and Lucas Numbers with Applications*, (John Wiley & Sons, New York).

Krutetskii, V. A. (1976). *The Psychology of Mathematical Abilities in School Children*, (University of Chicago Press, Chicago).

Lesh, R., Galbraith, P. L., Haines, C. R. and Hurford, A. (2010). (Eds), *Modeling Students' Mathematical Modeling Competencies: ICTMA 13*, (Springer, New York).

Lucas, E. (1891). *Théorie des Nombres* (Tome 1), (Gauthier-Villars et Fils, Paris).

Maddux, C. D. (1984). Educational microcomputing: the need for research, *Computers in the Schools*, 1(1), pp. 35-41.

Mason, J. and Pimm, D. (1984). Generic examples: seeing the general in the particular. *Educational Studies in Mathematics*, 15(3), pp. 277-289.

Matiyasevich, Yu. V. (1999). Formulas for prime numbers. In S. Tabachnikov (Ed.), *Kvant Selecta: Algebra and Analysis, II*, (The American Mathematical Society, Providence, RI) pp. 13-24.

Ministry of Education, Singapore. (2006). *Secondary Mathematics Syllabuses*, Curriculum Planning and Development Division: Author. Available at: http://www.moe.edu.sg/education/syllabuses/sciences/files/ maths-secondary.pdf.

Ministry of Education, Singapore. (2012). *N(T)-level Mathematics Teaching and Learning Syllabus*, Curriculum Planning and Development Division: Author. Available at: http://www.moe.gov.sg/education/syllabuses/sciences/files/ normal-technical-level-maths-2013.pdf.

Mishra, P. and Koehler, M. J. (2006). Technological pedagogical content knowledge: A framework for teacher knowledge, *The Teachers College Record*, 108(6), pp. 1017-1054.

Mollin, R. A. (1997). Prime-producing quadratics, *The American Mathematical Monthly*, 104(6), pp. 529-544.

Murray, M. (2004) *Teaching Mathematics Vocabulary in Context: Windows, Doors, and Secrete Passageways*, (Heinemann, Portsmouth, NH).

Musser, G. L. and Burger, W. F. (1997). *Mathematics for Elementary Teachers: A Contemporary Approach*, (Prentice Hall, Upper Saddle River, NJ).

National Council of Teachers of Mathematics. (1991). *Professional Standards for Teaching Mathematics*, (Author, Reston, VA).

National Council of Teachers of Mathematics. (2000). *Principles and Standards for School Mathematics*, (Author, Reston, VA).

National Curriculum Board. (2008). *National Mathematics Curriculum: Framing Paper*, (Author, Australia). Available at: http://www.ncb.org.au/verve/_resources/National_Mathematics_ Curriculum_-_Framing_Paper.pdf.

New York State Education Department. (1996). *Learning Standards for Mathematics, Science, and Technology*, (Author, Albany, NY).

New York State Education Department. (1998). *Mathematics Resource Guide with Core Curriculum*, (Author, Albany, NY).

Newman, D., Griffin, P. and Cole, M. (1989). *The Construction Zone*, (Cambridge University Press, Cambridge, MA).

Niess, M. L. (2005). Preparing teachers to teach science and mathematics with technology: developing a technology pedagogical content knowledge, *Teaching and Teacher Education*, 21(5), pp. 509-523.

Núñez, R. E., Edwards, L. D., and Matos, F. J. (1999). Embodied cognition as grounding for situatedness and context in mathematics education, *Educational Studies in Mathematics*, 39(1-3), pp. 45-65.

Nunokawa, K. (1995). Problem solving as modeling: A case of augmented-quotient division problem, *International Journal of Mathematical Education in Science and Technology*, 26(5), pp. 721-727.

Ontario Ministry of Education. (2005a). *The Ontario Curriculum, Grades 1-8, Mathematics (revised)* [on-line materials]. Available at: http://www.edu.gov.on.ca.

Ontario Ministry of Education. (2005b). *The Ontario Curriculum, Grades 9 and 10, Mathematics (revised)* [on-line materials]. Available at: http://www.edu.gov.on.ca.

Palatnik, A. and Koichu, B. (2014). What counts for being creative? A mathematically gifted students' perspective, In *Proceedings of the 8th International Mathematical Creativity and Giftedness*

Conference, (Institute for the Development of Gifted Education, Denver, CO) pp. 96-101. Available at: http://www.igmcg.org/images/proceedings/MCG-8-proceedings.pdf.

Pask, C. (1998). The monkey and coconuts puzzle: exploring mathematical approaches, *Teaching Mathematics and Its Applications*, 17(3), pp. 123-131.

Pegg, E., Jr. (2006). *Prime Generating Polynomials* [on-line materials]. Available at: http://www.mathpuzzle.com/MAA/48-Prime%20Generating%20Polynomials/mathgames_07_17_06.html.

Pollak H. O. (1970). The applications of mathematics, In Begle, E. G. (Ed.), *The Sixty-Ninth Yearbook of the National Society for the Study of Education*, (The National Society for the Study of Education, Chicago) pp. 311-334.

Pólya G. (1954). *Induction and Analogy in Mathematics*, (Princeton University Press, Princeton, NJ).

Pólya G. (1957). *How to Solve It*, (Doubleday, New York).

Pugalee, D. K. (2001). Writing, mathematics, and metacognition: looking for connections through students' work in mathematical problem solving, *School Science and Mathematics*, 101 (5), pp. 236-245.

Rakocevic, M. M. (2004). Further generalization of Golden Mean in relation to Euler's "divine" equation, *FME Transactions*, 32(2), pp. 95-98.

Roberts, J. M. and Westad, O. A. (2013). *The History of the World*, (Oxford University Press, New York).

Roberts, S. (1879). Problem 5446. *Mathematical Questions and Problems*, 30, p. 37.

Rohlin, V. A. (2013). *A Lecture about Teaching Mathematics to Non-Mathematicians*. Part I. Available at: http://mathfoolery.wordpress.com/2011/01/01/a-lecture-about-teaching-mathematics-to-non-mathematicians/.

Roy, R. (2011). *Sources in the Development of Mathematics: Infinite Series and Products from the Fifteenth to the Twenty-First Century*, (Cambridge University Press, Cambridge, MA).

Saaty, S. L. and Alexander, J. M. (1981). *Mathematical Models and Applications*, (Pergamon, Oxford, UK).

Schoenfeld, A. H. (1992). Learning to think mathematically: problem solving, metacognition, and sense making in mathematics, In D. A.

Grouws (Ed.), *Handbook of Research on Mathematics Teaching and Learning*, (Macmillan, New York) pp. 334-370.

Science Centre Singapore. (2010). *Mathematics: Everywhere & Everyday* [on-line materials], (Author, Singapore). Available at: http://www.science.edu.sg/exhibitions/Pages/Mathematics.aspx.

Shulman, L. S. (1986). Those who understand: knowledge growth in teaching, *Educational Researcher*, 15(2), pp. 4-14.

Shulman, L. S. (1987). Knowledge and teaching: foundations of the new reform, *Harvard Educational Review*, 57 (1), pp. 1-22.

Silverman, J. H. (1993). Taxicabs and sums of two cubes, *The American Mathematical Monthly*, 100(4), pp. 331-340.

Singh, P. (1985). The so-called Fibonacci numbers in ancient and medieval India, *Historia Mathematica*, 12(3), pp. 229-244.

Smith, D. E. (1959) *A Source Book in Mathematics, Vol. 2*, (Dover, New York).

Stakhov, A. P. (2004). Generalized golden sections and a new approach to the geometric definition of a number, *Ukrainian Mathematical Journal*, 56(8), pp. 1362-1370.

Steele, D. and Johanning, D. (2004). A schematic-theoretic view of problem solving and development of algebraic thinking, *Educational Studies in Mathematics*, 57(1), pp. 65-90.

Steuding, J. (2005). *Diophantine analysis*, (Chapman and, Hall/CRC, Boca Raton, FL).

Stewart, I. (1990). Change, In L. A. Steen (Ed.), *On the Shoulders of Giants: New Approaches to Numeracy*, (National Academies Press, Washington, D.C.) pp. 183-217.

Stockmeyer, P. K., Bateman, C. D., Clark, J. W., Eyster, C. R., Harrison, M. T., Loehr, N. A., Rodriguez, P. J. and Simmons, J. R. (1995). Exchanging disks in the Tower of Hanoi, *International Journal of Computer Mathematics*, 59(1-2), pp. 37-47.

Sugden S. (2005). Colour by numbers: solving algebraic equations without algebra, *Spreadsheets in Education*, 2(1), Article 6. Available at: http://epublications.bond.edu.au/ejsie/vol2/iss1/6.

Sugden, S. and Miller, D. (2010). Exploring the Fundamental Theorem of Arithmetic in Excel 2007, *Spreadsheets in Education*, 4(2), Article 2. Available at: http://epublications.bond.edu.au/ejsie/vol4/iss2/2.

Sutherland R. (1991). Some unanswered research questions on the teaching and learning of algebra, *For the Learning of Mathematics*, 11(3), pp. 40-46.

Sutherland, R. and Rojano, T. (1993). A spreadsheet approach to solving algebra problems, *Journal of Mathematical Behavior*, 12(4), pp. 353-83.

Szetela, W. (1999). Triangular numbers in problem solving, *Mathematics Teacher*, 92(9), pp. 820-824.

Tabach, M. and Friedlander, A. (2008). Understanding equivalence of symbolic expressions in a spreadsheet-based environment, *International Journal of Computers for Mathematical Learning*, 13(1), pp. 27-46.

Takahashi, A., Watanabe, T., Yoshida, M., and McDougal, T. (2004). *Elementary School Teaching Guide for the Japanese Course of Study: Arithmetic (Grade 1-6)*, (Global Education Resources, Madison, NJ).

Takahashi, A., Watanabe, T., Yoshida, M., and McDougal, T. (2006). *Lower Secondary School Teaching Guide for the Japanese Course of Study: Mathematics (Grade 7-9)*, (Global Education Resources, Madison, NJ).

Tall, D., Gray, E., Ali, M. B., Crowley, L., DeMarois, P., McGowen, M., Pitta, D., Pinto, M., Thomas, M., and Yusof, Y. (2001). Symbols and the bifurcation between procedural and conceptual thinking, *Canadian Journal of Science, Mathematics, and Technology Education*, 1(1), pp. 80-104.

Trigg, C. W. (1974). All three digit integers lead to ... , *The Mathematics Teacher*, 67(1), pp. 41-45.

Uspensky, J. V. and Heaslet, M. A. (1939). *Elementary Number Theory*, (McGraw Hill, New York).

Van de Walle, J. A., Karp, K. S., and Bay-Williams, J. M. (2013). *Elementary and Middle School Mathematics: Teaching Developmentally*, 8th edition, (Pearson, New York).

Van der Waerden, B. L. (1961). *Science Awakening*, (Oxford University Press, New York).

Vergnaud, G. (1982). Cognitive and development psychology research on mathematics education: some theoretical and methodological issues, *For the Learning of Mathematics*, 3 (2), pp. 31-41.

Vygotsky, L. S. (1962). *Thought and Language*, (The MIT Press, Cambridge, MA).

Vygotsky, L. S. (1987). Thinking and speech, In R. W. Rieber and A. S. Carton (Eds), *The Collected Works of L. S. Vygotsky, vol. 1*, (Plenum Press, New York) pp. 39-285.

Walsh, C. M. (1927-1928). Fermat's note XLV. *Annals of Mathematics*, 29(1/4), pp. 412-432.

Watson, A. and Mason, J. (2005). *Mathematics as a Constructive Activity: Learners Generating Examples*, (Lawrence Erlbaum Associates, Hillsdale, NJ).

Weisstein, E. W. (1999). *The CRC Concise Encyclopedia of Mathematics*, (CRC Press, Boca Raton, FL).

Western and Northern Canadian Protocol. (2008). *The Common Curriculum Framework for Grades 10-12 Mathematics* [on-line materials]. Available at: http://www.bced.gov.bc.ca/irp/pdfs/mathematics/WNCPmath1012/2008math1012wncp_ccf.pdf.

Williams, S. W. (2002). Million-buck problems, *Mathematical Intelligencer*, 24(3), pp. 17-20.

Yerushalmy, M., Chazan, D., and Gordon, M. (1993). Posing problems: One aspect of bringing inquiry into classrooms, In J. L. Schwartz, M. Yerushalmy, and B. Wilson (Eds), *The Geometric Supposer: What it is a Case of?* (pp. 117-142), Hillsdale, NJ: Lawrence Erlbaum Associates.

Index

4-D Modeling, 85

Abramovich, S., 10, 127, 177, 199-201, 239, 243
absolute reference, 6, 113
addition table, 30, 253
Advisory Committee on Mathematics Education, vi, 78, 160, 177, 179, 180, 285
Agent-Consumer-Amplifier Framework, 9
Alfors, L., 180
algebraic inequalities, 177
arithmetic mean-geometric mean, 189
arithmetic sequence, 5, 8, 17, 22, 95, 199, 200, 205, 211, 252
attractor, 172
automatic skills, 24

Babylonian algebra, 181
Babylonian mathematics, 177, 190
Bachet, C. G., 146
Baker, J. E., 199
Ball, D. L., 197
base clause, 32
base ten blocks, 161
Becker, J. P., 262
Beth, E. W., 113
Binet, J. P. M., 239
Blum, W., 180, 201
Brown, S. I., 180
Buniakovski, V., 99
Burkert, W., 181

Cai, J., 77
checkerboard, 30
circular reference, 13, 14, 75, 76, 81, 100, 106, 129, 191, 192
closed formula, 207
Cockcroft Report, 10, 109

collateral learning, 145
Common Core State Standards, 33, 37, 69, 109, 190
computing complexity, 123
Comtet, L., 70
conceptual understanding, 180
conditional formatting, 32, 38
Conference Board of the Mathematical Sciences, v, 29, 158, 177, 178, 189, 201, 232, 285
conscious awareness, 2
contextual coherence, 78
converging differences, 172
Conway, J. H., 103
Cooke, R., 133
COUNTIF function, 22
cube attractors, 262
Cuoco, A., 152, 179
cyclic behavior, 24, 26
cyclic sequences, 215

denumerant, 70, 73
Department for Education, viii, 30, 109, 145, 157, 201, 207, 287
descriptive name, 5
Dickson, L. E., 227
didactical coherence, 78
difference equation, 2, 206, 232, 233, 237, 239
Diophantine equation, 10
Diophantus, 10, 146
direct proof, 191
Dirichlet, P. G. L., 95
dividend, 128
divisor, 128
Doerr, H. M., 180

empirical induction, 31
Engel, A., 201
environment, 2

281

Eratosthenes, 89
Euclid, 88, 125, 133, 142
Euclidean algorithm, 122, 126-128
Euler phi function, 129, 149
Euler, L., 88, 138, 227
Euler's factorization method, 104, 259

face value, 157
Fermat primes, 103
Fermat, P., 103, 138
Fermat's Last Theorem, 146
Fibonacci numbers, 124, 128, 231, 232, 237, 241
Fibonacci sieve, 235, 237, 239, 244
Fibonacci, L., 231
Fibonacci-like sequence, 232
formal demonstration, 22, 114
function FLOOR(x, 1), 11
function IF, 8, 137, 147, 169
function INDEX, 12
function INT(x), 11
function LEN(x), 166
function LOOKUP, 12
function OR, 71

gaps between primes, 97
Gauss, C. F., 87, 103, 146
generalized golden ratios, 243
Getzels, J. W., 77
gnomon, 32, 34, 38
Golden Ratio, 241, 242, 246, 249
greatest common divisor, 116
Gruenberger, F., 171

Harrison, J., 194
hidden mathematics curriculum, 95
Hoyles, C., 199, 200, 201
human-computer interaction, 19

inductive transfer, 32
inequality, 189

Kaprekar, D., 172
Kline, M., 181, 188
Krutetskii, V. A., 77

Lamé, G., 126
Law of Cosines, 150
Legendre, A.-M., 88
Leonov, G. A., 239, 243
Lesh, R., 180
Lucas numbers, 232
Lucas, E., 232

Maddux, C. D., 197
manipulative-computational environment, 109
Maple, vi, 18, 19, 101, 227, 240, 249, 285
Mason, J., 35, 77
mathematical induction proof, 19, 208, 242
mathematical modeling, 177
Matiyasevich, Yu. V., 95
measurement model for division, 158
method of mathematical induction, 249, 252
m-gonal numbers, 224
Ministry of Education, Singapore, vi, 29, 48, 51, 109, 157, 167, 177, 178, 180, 196, 197, 286
MOD function, 21
multiplication table, 35, 57, 61, 254

National Council of Teachers of Mathematics, 35, 178, 198, 201
National Curriculum Board, vi, 157, 171, 194, 286
necessary and sufficient condition, 208
nested IF function, 27
New York State Education Department, 70, 79, 158
non-authoritative pedagogy, 110
number theory, 87

numerical coherence, 78 .
numerical evidence, 20, 22, 113, 116, 135, 190, 251
numerical evidence, 6
Nunokawa, K., 180

On-line Encyclopedia of Integer Sequences, 174
Ontario Ministry of Education, vii, 29, 109, 117, 151, 157, 182, 207, 287
ostensive definition, 3, 165, 232

palindrome, 158
Palindromic Number Conjecture, 171, 172
partition, 85
pedagogical coherence, 78
Pegg, E., 100, 101
pentagonal number, 34, 265, 267
percentage problems, 109
Pimm, D., 35
place value, 157
Plutarch, 194
Pólya, G., 17, 84, 135, 208
polygonal numbers, 88, 214
prime numbers, 87, 99, 101, 103, 124, 130
prime-producing polynomials, 98
prime-producing quadratics, 100
primitive Pythagorean triples, 133
problem posing, 77, 78, 82, 152
problem solving, 78, 180
proof assistant technology, 194
proof by contradiction, 184
Pythagoras, 133
Pythagorean equation, 133
Pythagorean triangles with equal areas, 144
Pythagorean triple, 133, 140, 142, 149

quotient, 128

recreational mathematics, 157
recursive definition, 2, 6, 207, 211
recursive formula, 214
relative reference, 6, 76
remainder, 128

Saaty, S. L., 180
Schoenfeld, A. H., 134
Science Centre Singapore, 231
scroll bar, 4
seed value, 7, 24, 26, 212, 214
self-reproducing endpoint, 173
shallow diagonals, 47
Shulman, L. S., 197
sieve of Eratosthenes, 91, 130
slider, 4
slope, 215
Smith, D. E., 146
square numbers, 210
square-like numbers, 221, 223
Stewart, I., 146
structured, 180
Sugden, S., 199
sums of two squared integers, 135
Sutherland, R., 199
symbolic computations, 17

Takahashi, A., vi, 51, 73, 116, 286
Tall, D., 20, 37
technological pedagogical content knowledge, 197
technology-immune/technology-enabled problem, 84
three-dimensional modeling, 70
TITE problem, 69
triangular number, 37, 205, 207
triangular squares, 98, 226-228
triangular-like numbers, 217, 223
trigonometry, 150
twin primes, 95
two-dimensional modeling, 29, 91

Van der Waerden, B. L., 190, 194

Vergnaud, G., 122, 149

Vygotsky, L. S., 2, 165

Walsh, C. M., 145

Watson, A., 77

Weisstein, E. W., 143, 171

Western and Northern Canadian Protocol, 69, 115, 157, 177, 178, 184

Wiles, A., 146

Williams, S. W., 171

Wolfram Alpha, vi, 17, 56, 72, 143, 227, 249, 285

Yerushalmy, M., 77

Printed in the United States
By Bookmasters